Montessori Math Workbook

A HANDS-ON APPROACH TO EARLY MATHEMATICS

PRIMARY BOOK 2

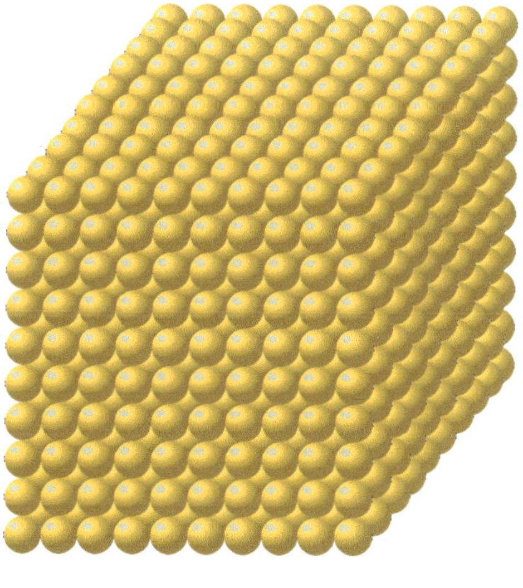

by Katie Key
montessoriforhomeschoolers.com

ISBN: 978-1-7353980-0-6
Library of Congress Control Number: 1-9002793911

Montessori Math Workbook - Primary Book 2

Graphics Used:
-Bead Clipart is from Bee Creative Cliparts.
- Letter Formation Font is from RamonaM Graphics.

montessoriforhomeschoolers.com
raisingkingdomwarriors.com

Policy Page

By purchasing and/or downloading this workbook, you are agreeing to the following terms and conditions:

Family License:

1. This Workbook is to be used only by the immediate family of the purchaser. You may not distribute by giving or selling this workbook in any format to any person outside of your immediate family.
2. You may print as many copies for your immediate family as you would like.
3. Please do not use this workbook for multiple children in a classroom or co-op. If you need a set of physical Workbooks for your class or a license to use the digital version of this Workbook for your class, please email me directly at montessoriforhomeschoolers@gmail.com for bulk pricing.
4. Please do not sell or re-use any or all parts of this workbook for profit or otherwise.

Classroom License:

1. By purchasing a digital classroom license, you have permission to make copies for a class of up to 30 students for **one year**.
2. You may print as many copies for your **single classroom** to be used by up to 30 students in a single class for one year.
3. You may distribute the copies digitally, but please include this page with your distribution, and let your students' parents know that they cannot use or distribute this product in any way for any purposes other than completing the assignment for your class.
4. If you need a set of physical Workbooks for your class or to upgrade to a license to use the digital version of this Workbook for your class, please email me directly at montessoriforhomeschoolers@gmail.com for bulk pricing.
5. Please do not sell or re-use any or all parts of this workbook for profit or otherwise.

Graphics Used:
- Bead Clipart is from Bee Creative Cliparts.
- Letter Formation Font is from RamonaM Graphics.

I trust that you will adhere to these policies, as it helps support a family like yours! I have worked very hard to create amazing resources for my family and yours, and I want to thank you for doing your part to support my small business!

Dear Parent/Teacher,

Thank you for purchasing the Montessori Math Workbook - Primary Book 2! I am amazed at the response to the first workbook, and I know that God is using this work to help parents and teachers reach little minds, preparing them for life, goodness, and beauty, while equipping them with the mathematical knowledge to grow into their full potentials and beyond. When Jesus walked the earth, He told His disciples to "Let the little children come to me, and do not hinder them, for the kingdom of heaven belongs to such as these" (Matthew 19:14). By spending your time teaching one of God's precious little children, you are showing him or her what it is to *love* like Jesus.

This is the *second* Workbook in an expanding collection of Montessori Math Workbooks! This workbook is written for the child who is ready to move on from the initial numbers and counting exercises and venture to the Decimal System, Golden Beads, and Squaring Chains.

The math work contained in this volume stays within the *concrete* principles. Your child will learn how one to four digit numbers are made and then progress to manipulating those large numbers using physical materials found in the workbook through the four operations: addition, subtraction, multiplication, and division.

This is not a "Definitive Guide to Montessori Math," rather it is a paper solution for the parent or teacher looking for an affordable, easy way to incorporate Montessori math at home or in the classroom.

You can use this workbook as a **supplement** to traditional Montessori math work or as a **stand-alone** math curriculum.

I welcome and encourage feedback! I am human, so if there are errors, please do not hesitate to let me know. If you see a way that the workbook can be improved, I welcome those comments as well. We also can *all* benefit from encouragement, so I humbly ask for you to write me your praises so I can be encouraged, and so I can share them on my sales page if you give me permission.

Enjoy this journey with your child!

In Christ,

Katie Key
montessoriforhomeschoolers@gmail.com

Math Objectives Covered

Note: This workbook is an "Open and Go" Math Curriculum. Grab your pencil, crayons, scissors, and glue, and you can do any lesson in here. Many of the activities require cut-out materials so your child can physically manipulate the materials, so while everything is contained within the workbook, you may need a larger workspace to complete some of the activities.

In Primary Book 2, your child will be exposed to **Squares of Numbers 1-10**, **Skip Counting**, **the Decimal System**, and **Operations with Golden Beads**. There are also 60 days worth of "Review" questions that will review previously acquired skills.

Children as young as four years old will be adding, subtracting, multiplying, and dividing four digit numbers by the end of this book.

Celebrate your child's accomplishment in completing this workbook by presenting them with the special letter found on the very last page. You will give your child specific praise for his or her hard work, and you can record the date completed for your records.

Support Website

My goal in writing all of the directions and visual instructions for this workbook was to make it as "stand-alone" as possible for you! That being said, I know that some of us learn better (and more quickly) when we see someone do the exercise in front of us. So, I created some **instructional videos to supplement the on-paper directions**!

Go to: **https://montessoriforhomeschoolers.com/pages/math-book-2** for video instructions on some of the lessons, as well as any other supporting documents that may be released post-print.

Note: I will also post links to the physical Montessori materials that coincide with this workbook, should you choose to incorporate them into your learning environment. The materials are *not required* for a successful Montessori math experience using this workbook! Some families just prefer to use them in addition to the workbook.

TABLE OF CONTENTS

TABLE OF CONTENTS

Prep, Tips, Role Badges, and More

There are many activities in this workbook that require cutting, coloring, and writing numerals. As your child's parent and/or teacher, you know the capabilities of your individual child. In order to prevent some unnecessary frustration on your part and the part of your child, I wanted to offer some suggestions. Take what works, and leave what doesn't!

- Pre-cut the activity materials ahead of time, unless cutting is directly a part of the activity. Most activities have a moving component, and you will want evenly, nicely cut materials to work with!
- There are envelopes and "storage" pages for bead chains, arrows, and Golden Beads. You can prep those ahead of time so they are ready to be used by your child when the time comes.
- If your child is not a practiced writer, she may become fatigued when having to write numerals. Keep those activities short, or be the "Scribe" for your child. I have included badges that you can place in front of you and your child while you are working. Invite your child to decorate them! This takes some pressure off and keeps the emphasis where it belongs - learning math!
- If you have physical Montessori materials, always present with those first if you can, and use the activities in the workbook as follow-up.

There are icons on most lessons, so you know what is to be expected.

 Writing

 Coloring

 Hands-On Component (Moving objects around on a table surface)

 Movement (More than just moving things on a table; your child will likely need to get up!)

 Cut and Paste Activity

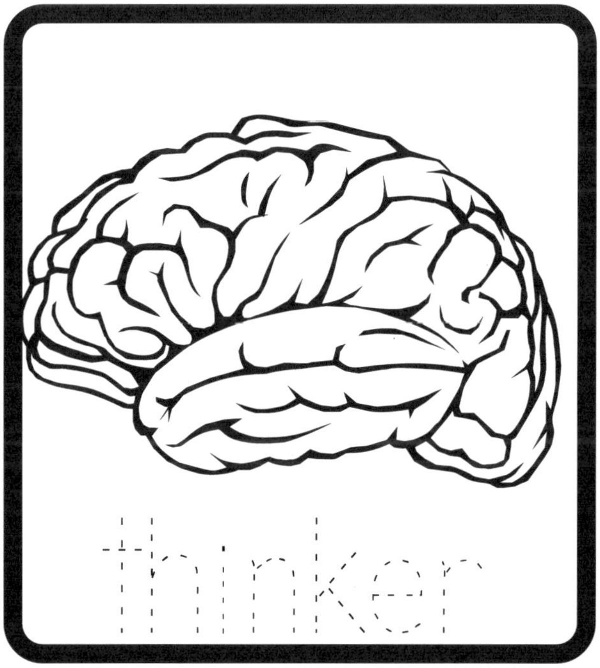

Squaring Chains

Directions:
When we show our children how to square chains, we give them an introduction to squaring, as well as to skip counting. Skip counting is a stepping stone to later multiplication.

First, you will give lessons on each of the chains. The lessons are broken up into three squares at a time, and we end with a review of squaring ten.

For each lesson, do the following:

1. Prep materials: cut out the square, the individual bead bars, and the colored bead arrows that you need for the lesson. Gather clear tape and glue! Place everything in a neat space that you can reach as you work. Lay out the arrows in numerical order.
2. First, place the cut-apart bead bars end-to-end vertically. Place the square of the beads at the top of the vertical chain of beads.
3. Now, carefully place the cut-apart bead bars directly on top of the square of the beads, covering up the square, so you show your child that the amount of beads is exactly the same. (No need to say anything, just place the beads!) Pause and observe the beads.
4. Lay out the beads vertically again, leaving the square at the top of the chain.
5. Tape the beads together to make a chain.
6. Count each individual bead, starting at the top, and place the colored arrows at their proper spots when you reach the number on each arrow.
7. Fold up the bead chains and keep them in their "envelope."
8. Gather the arrows and keep them in their "envelope." Do the same for the bead squares.
9. After completing each chain, keep a record of your child's progress and completion of each bead chain by filling in (or guiding your child to fill in) the **Record of Squares Worksheet.***

*NOTE: Completing all of the squaring chain exercises is not a requirement to move on to the Decimal System (Golden Bead Material) exercises in the next section. Your child can always come back and complete more bead chain squaring when he or she shows interest.

Lessons Progression Sample:
Day 1: 4's, 5's, 6's.
Day 2: 3's, 7's
Day 3: 1's, 2's, 8's
Day 4: 9's, 10's

Record of Squares

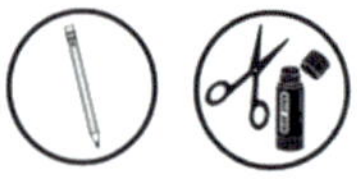

Directions:

Use the "Record of Squares" Worksheet to keep track of and review the Short Bead Chains that your child has completed.

To fill in the numbers on the bead chain arrows, use the physical, laid-out bead chain as a guide. Invite your child to copy the numbers from the physical arrows onto the arrows on the worksheet. Alternatively, you could ask your child to read the numbers to you and you could copy them. This will depend on your child's ability.

Then, cut and paste the small bead squares into the corresponding colored boxes at the end of each bead chain.

For verification, invite your child to count the beads again.

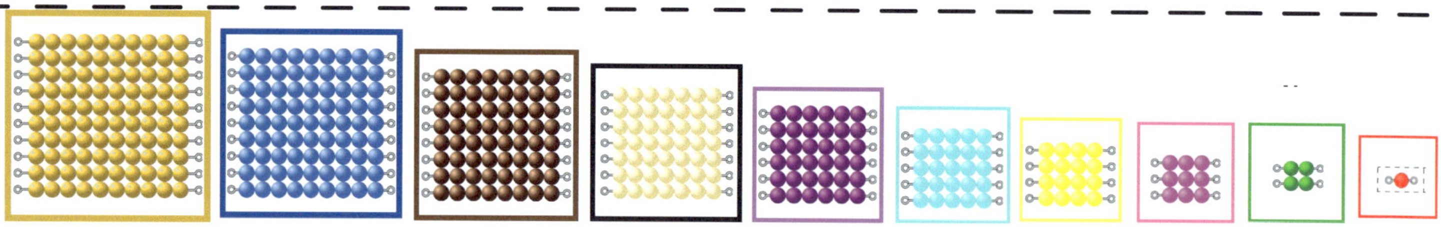

Record of Squares

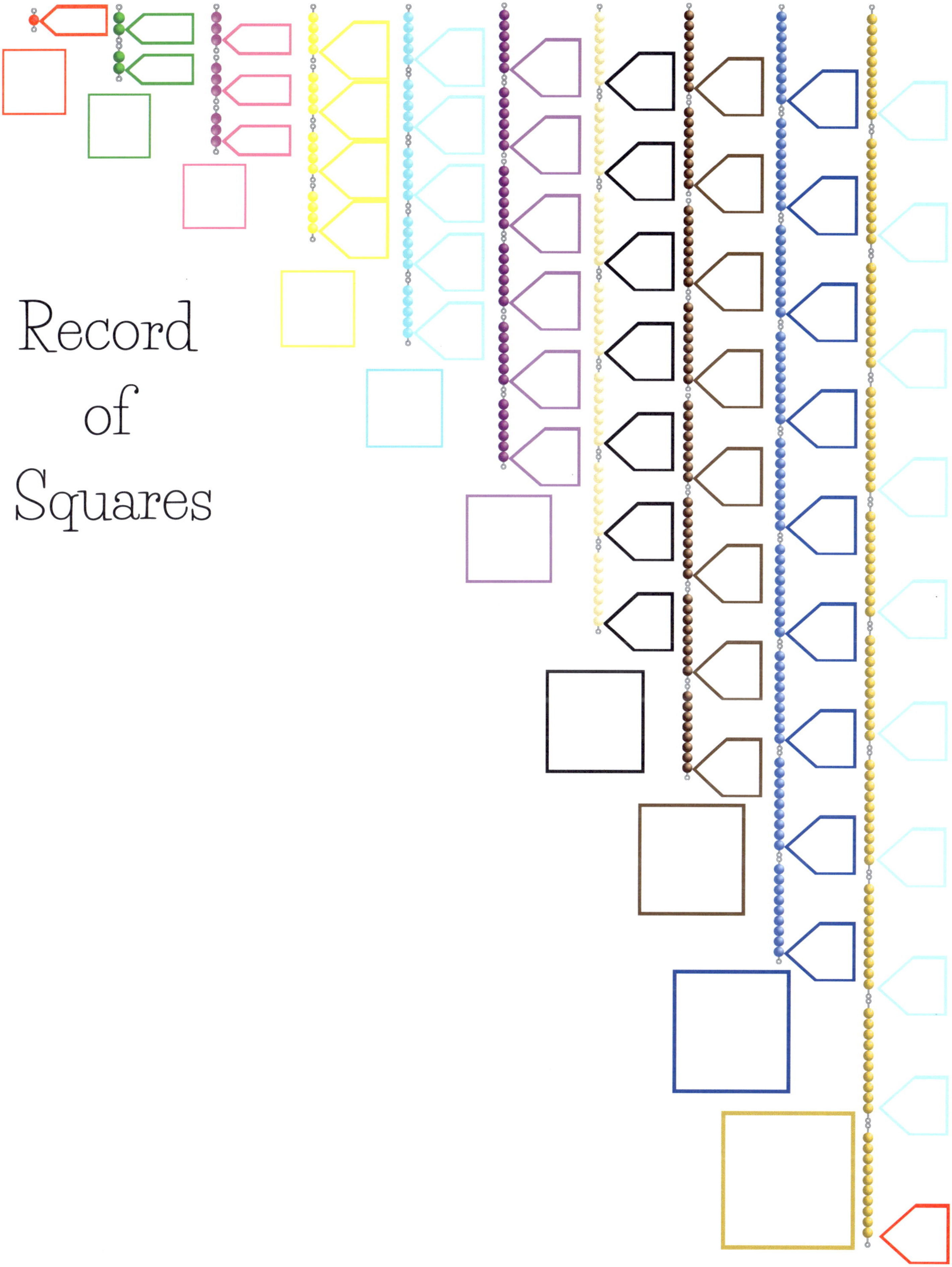

Bead Organizer

Use this organizer next to your workbook to organize your beads when working with the bead bars!

Bead Chain Arrows Organizer

Use this organizer next to your workbook to organize your bead arrows when working with the short bead chains!

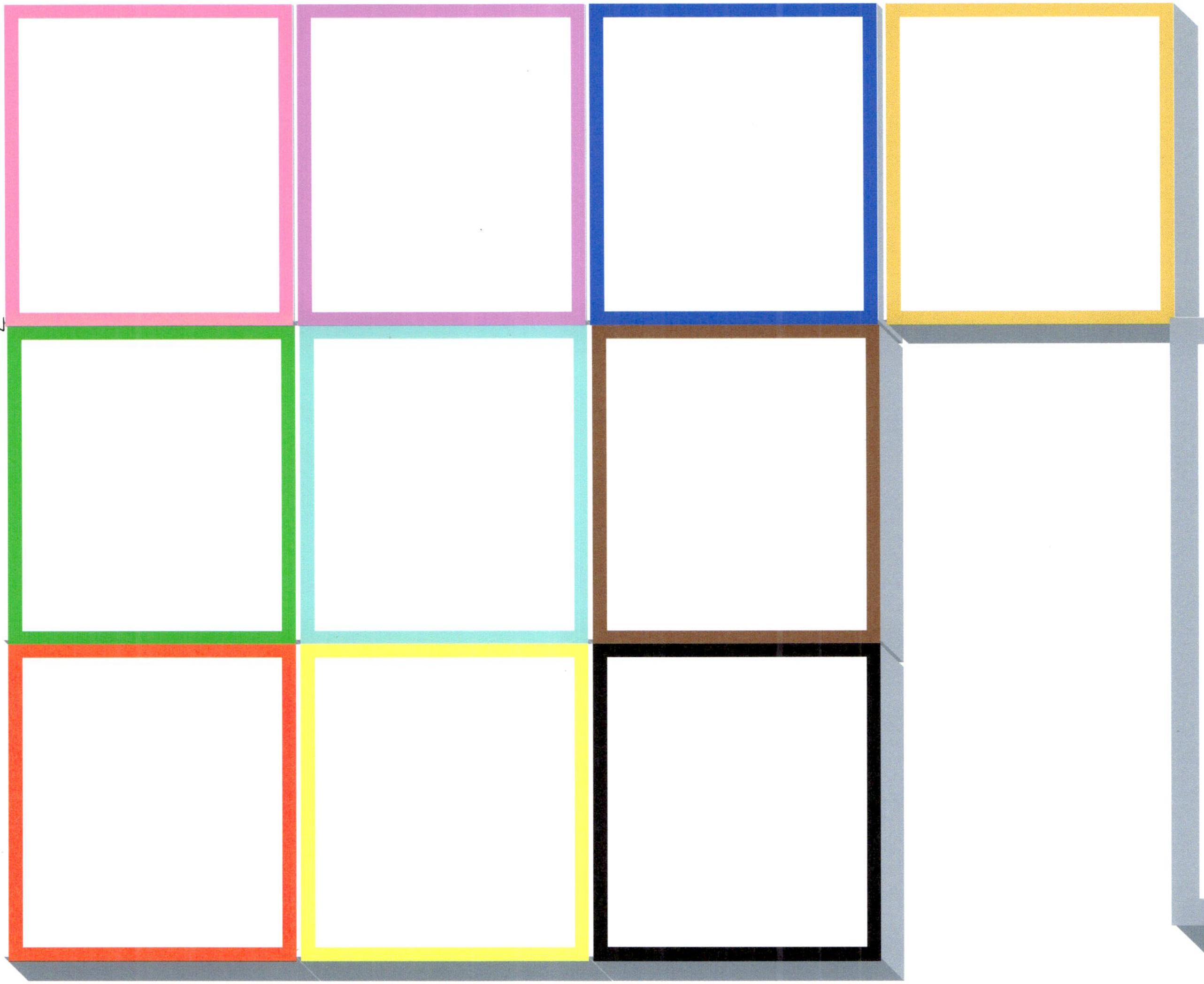

Bead Squares

Use this space to keep all of the bead squares.

Envelopes for Beads

Cut these apart on the outside square. Fold "in" along the inside lines. You should see the colors along the outside edges when you fold. Glue the rectangle middles onto the corresponding spots on your "Bead Chain Arrows " Organizer. Secure the bottom and side flaps with tape or glue. Use sticky tack or another removable adhesive to secure the top down when it is storing arrows.

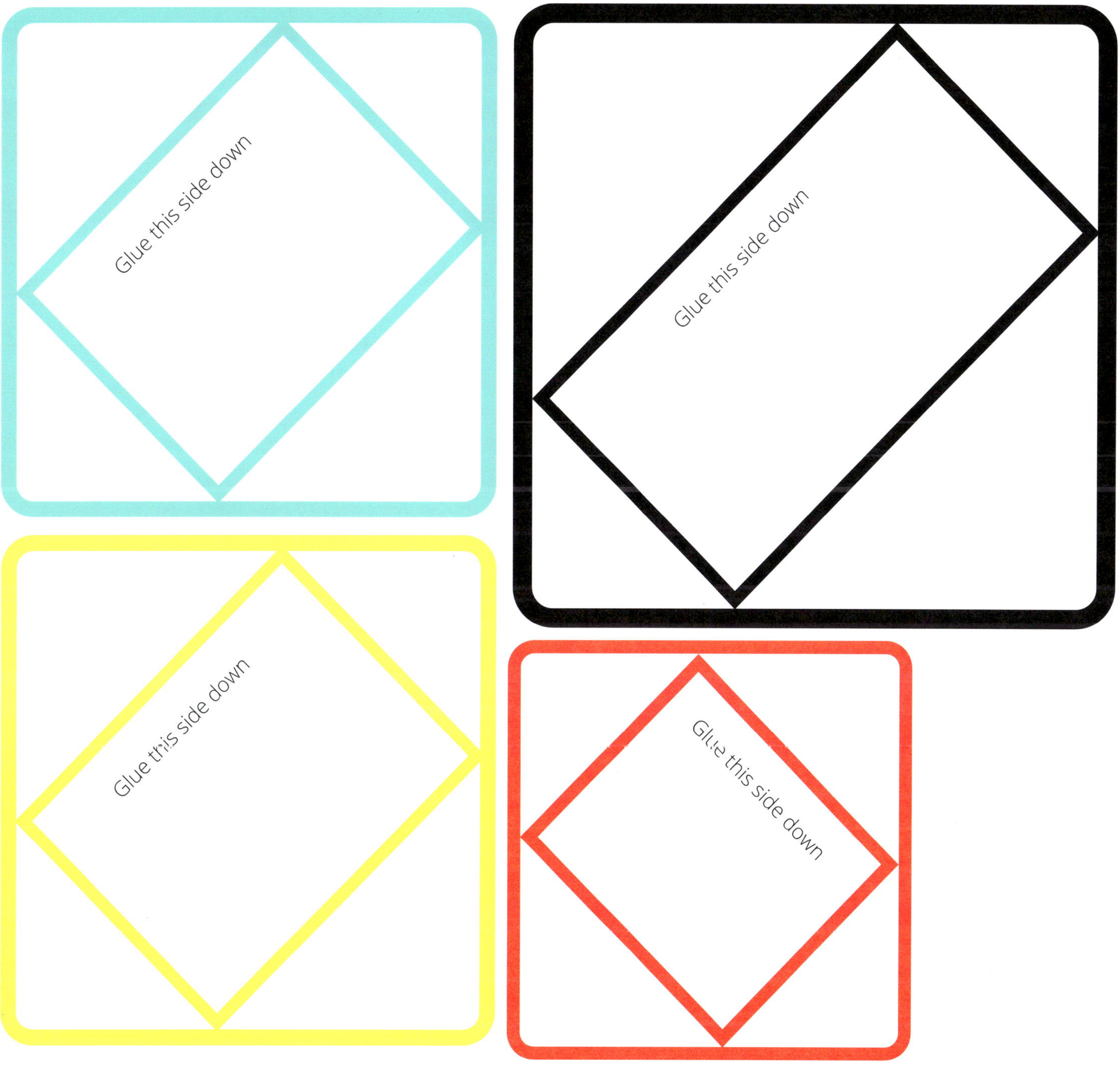

Glue this side down

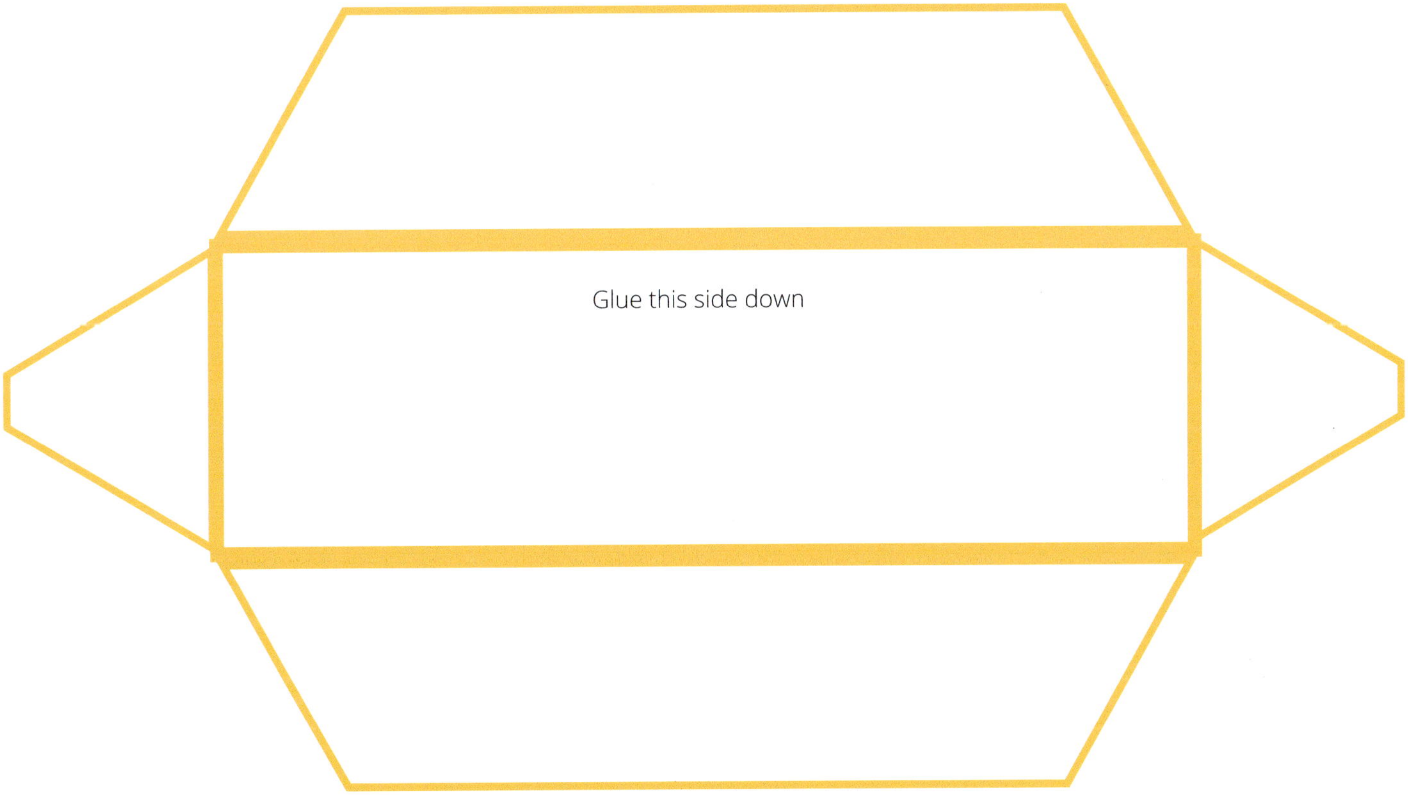
Glue this side down

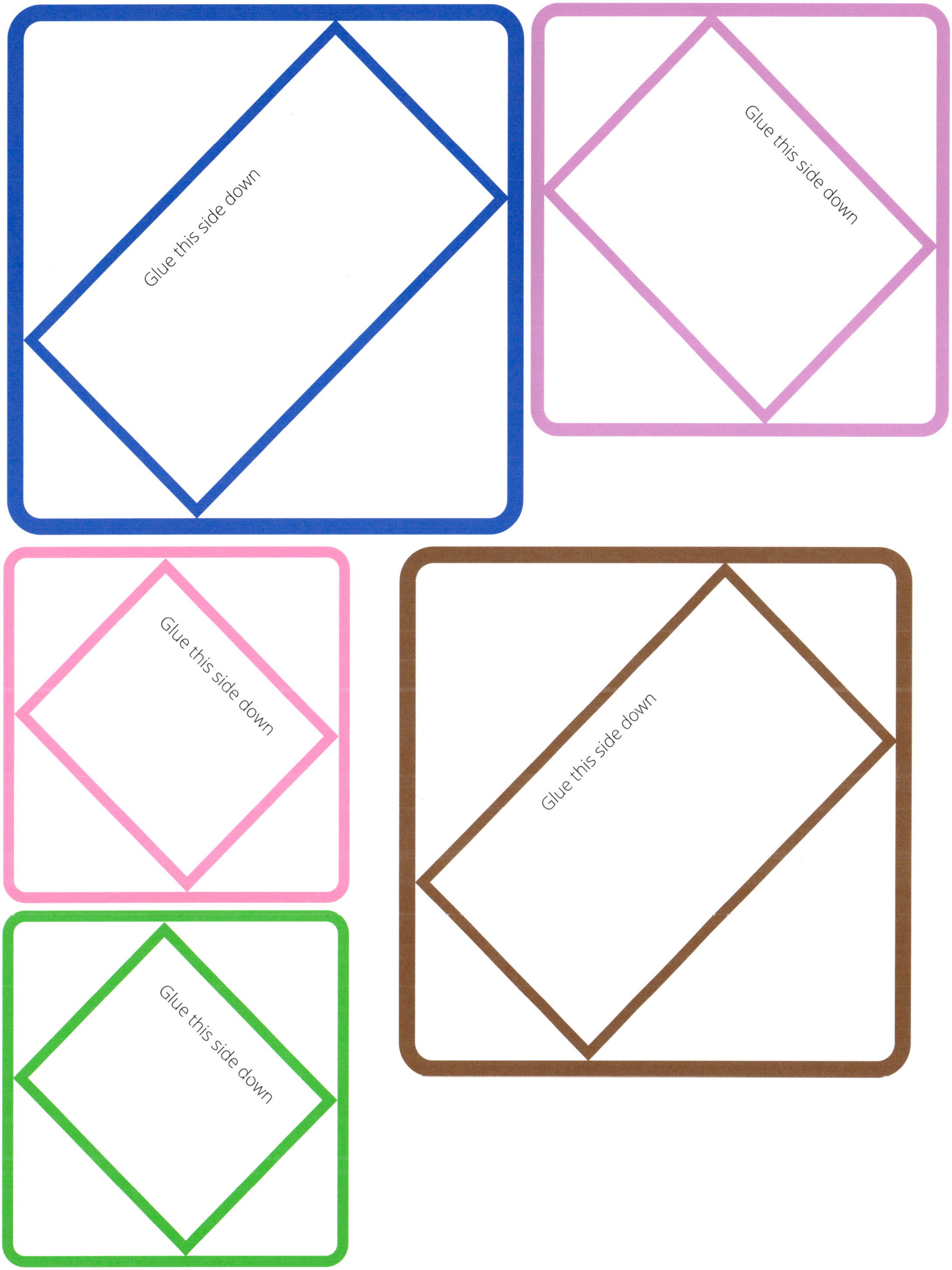
Glue this side down
Glue this side down
Glue this side down
Glue this side down
Glue this side down

Envelopes for Arrows

Cut these apart on the outside square. Fold "in" along the inside lines. You should see the colors along the outside edges when you fold. Glue the rectangle middles onto the corresponding spots on your "Bead Chain Arrows " Organizer. Secure the bottom and side flaps with tape or glue. Use sticky tack or another removable adhesive to secure the top down when it is storing arrows.

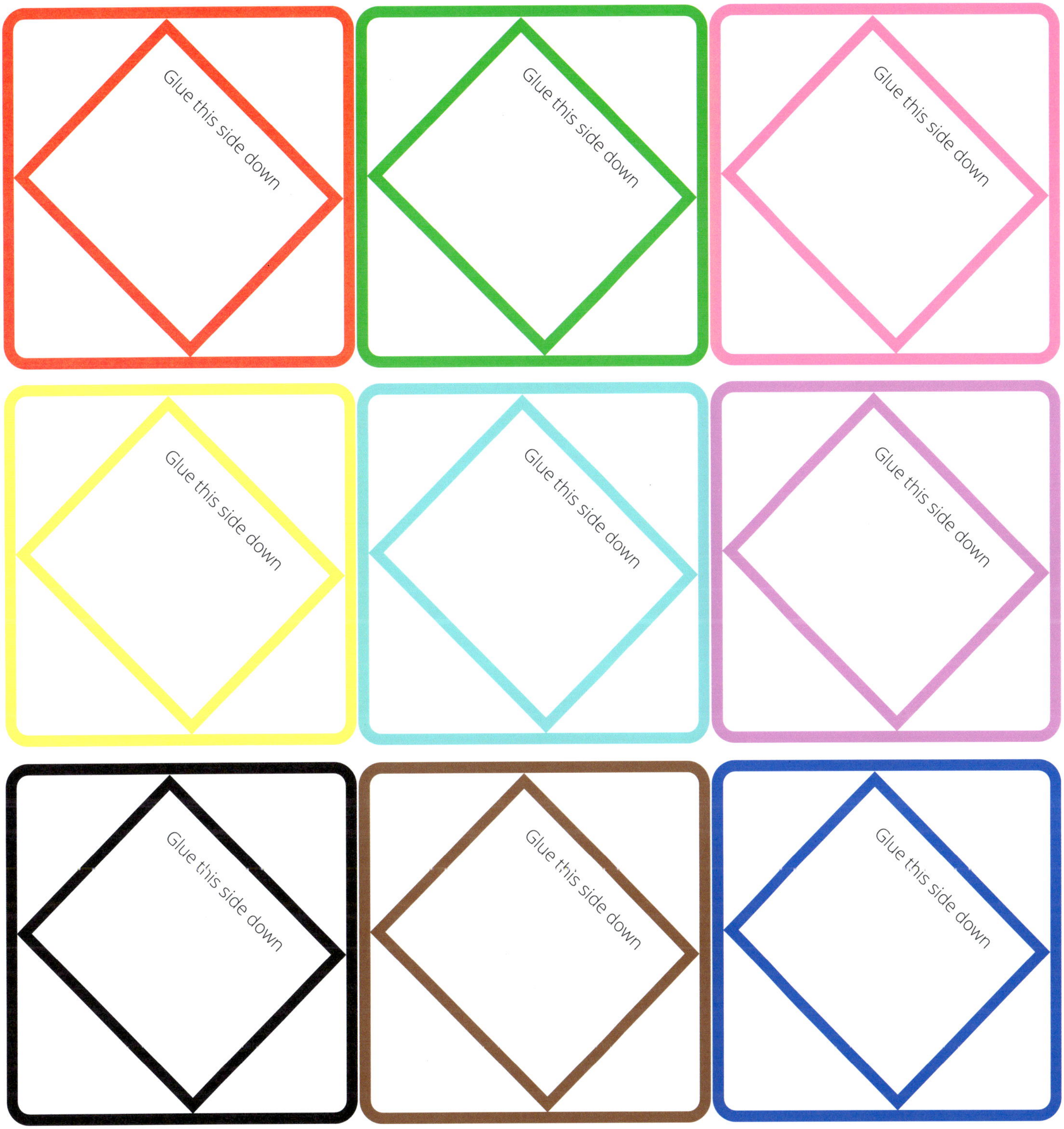

Short Bead Chain Arrows

9 18 27 36 45 54 63 72 81

7 14 21 28 35 42 49

5 10 15 20 25

1

2 4

3 6 9

6 12 18 24 30 36

4 8 12 16

8 16 24 32 40 48 56 64

10 20 30 40 50 60 70

80 90

100

Bead Chain Arrow
Envelope for Tens:

Beads for Short Bead Chains

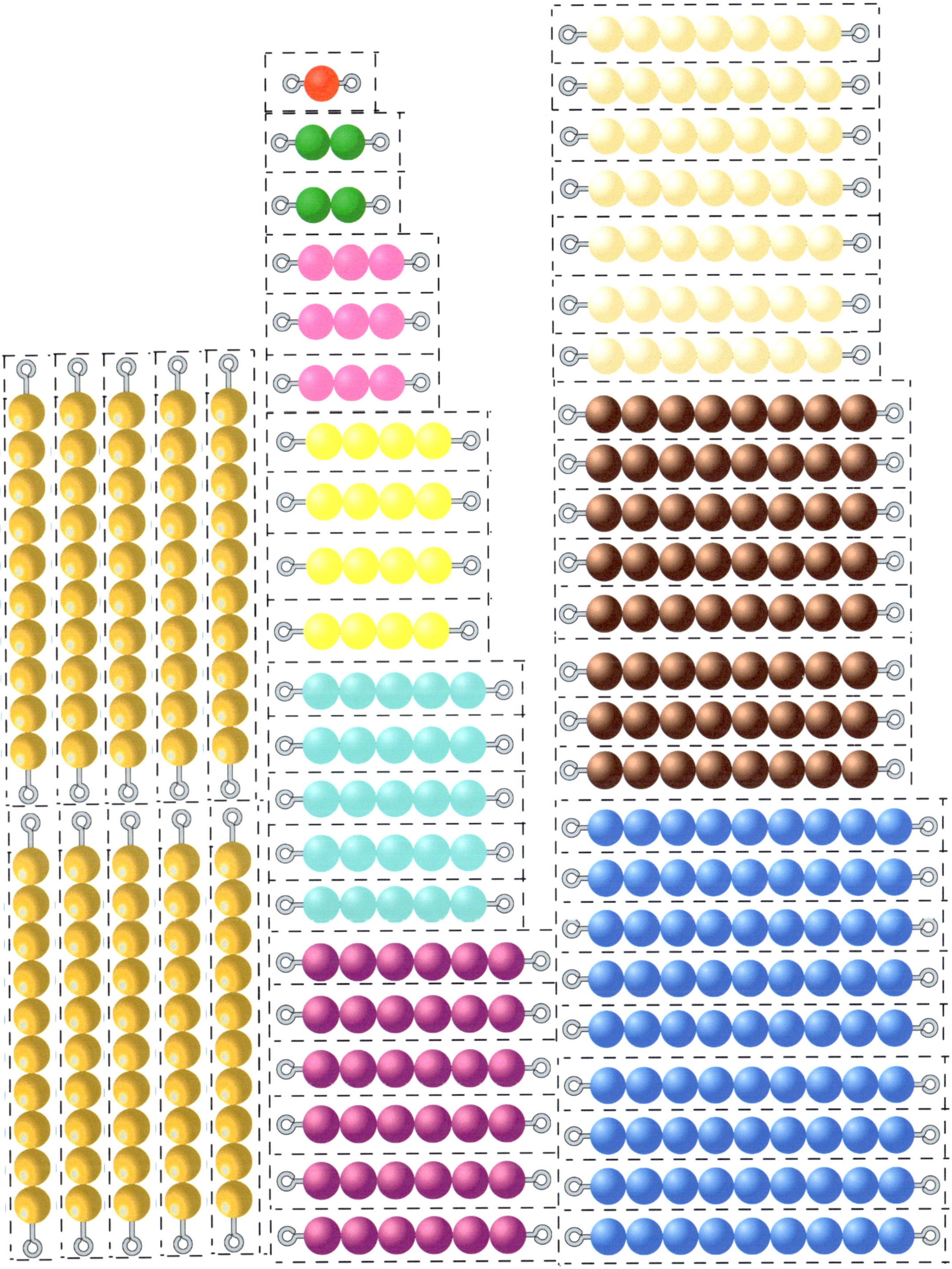

Bead Squares for Short Bead Chain Exercises

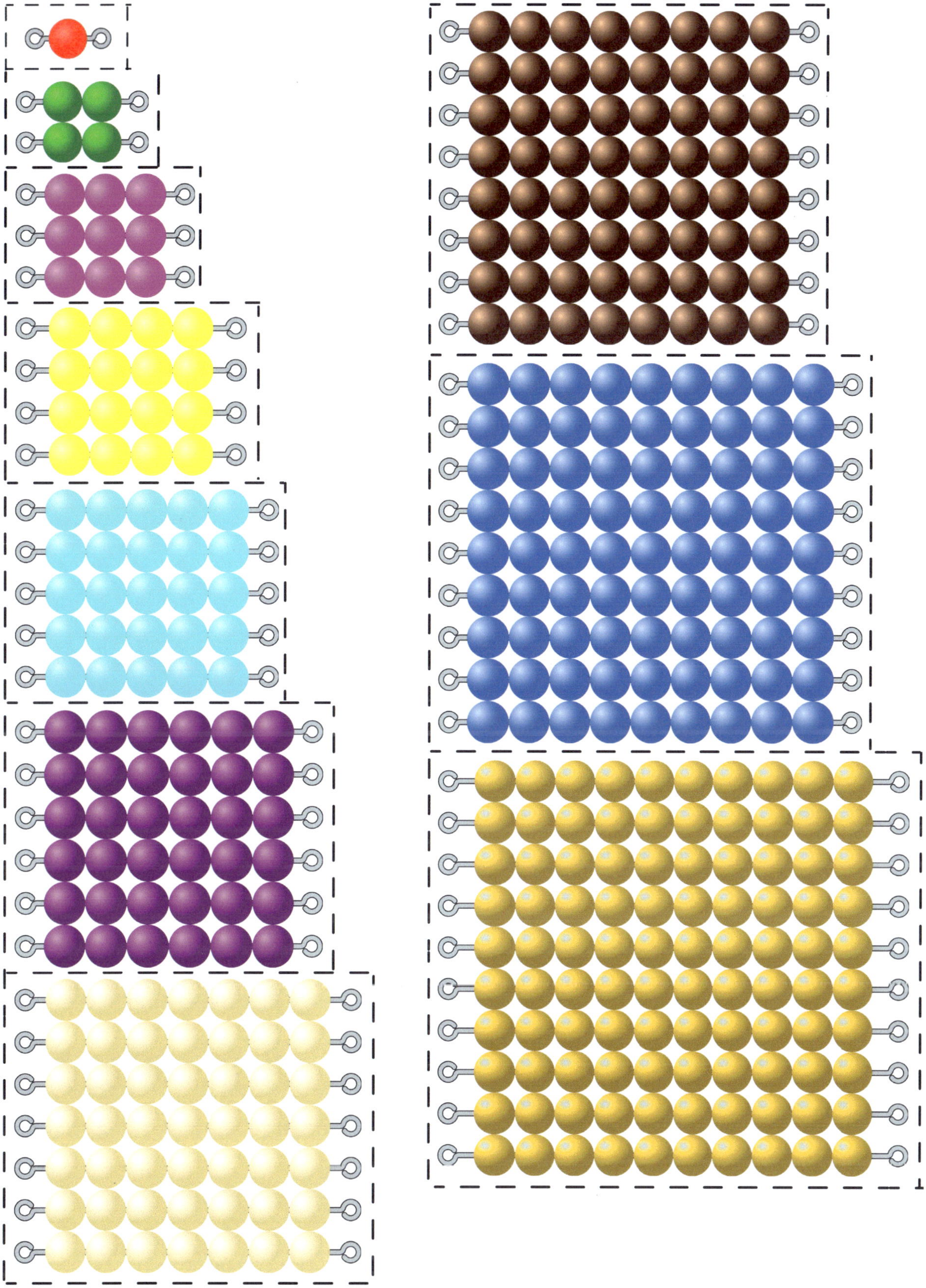

Trace and Match Short Bead Chains

Directions: Trace the numbers with a colored pencil that matches the color of the beads. Then draw a line from the tip of the arrow to the corresponding bead.

1

2

4

3

6

9

4

8

12

16

5

10

15

20

25

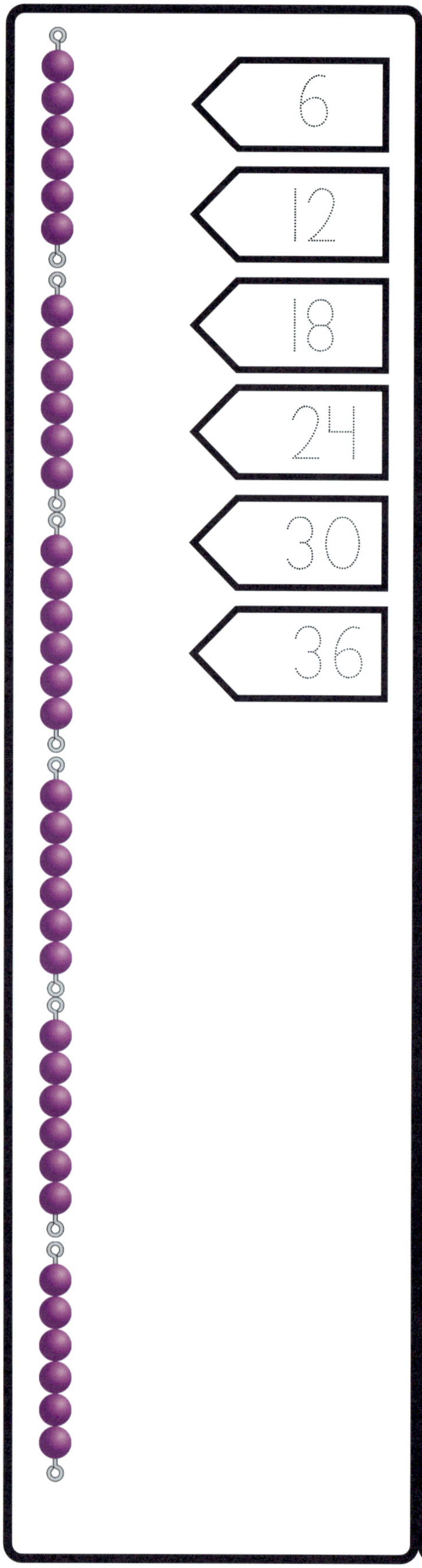
6
12
18
24
30
36

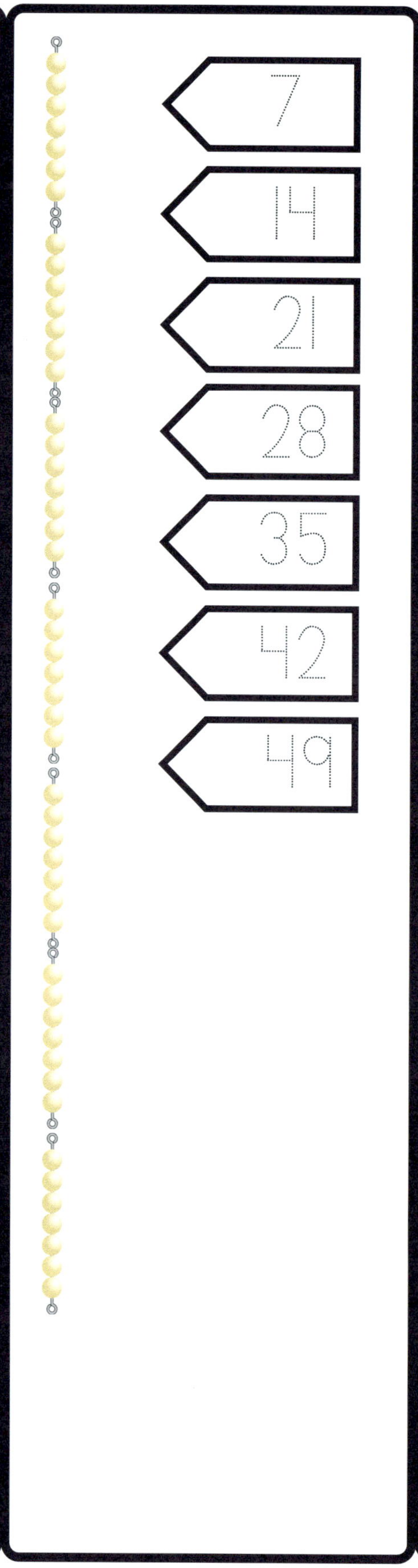
7
14
21
28
35
42
49

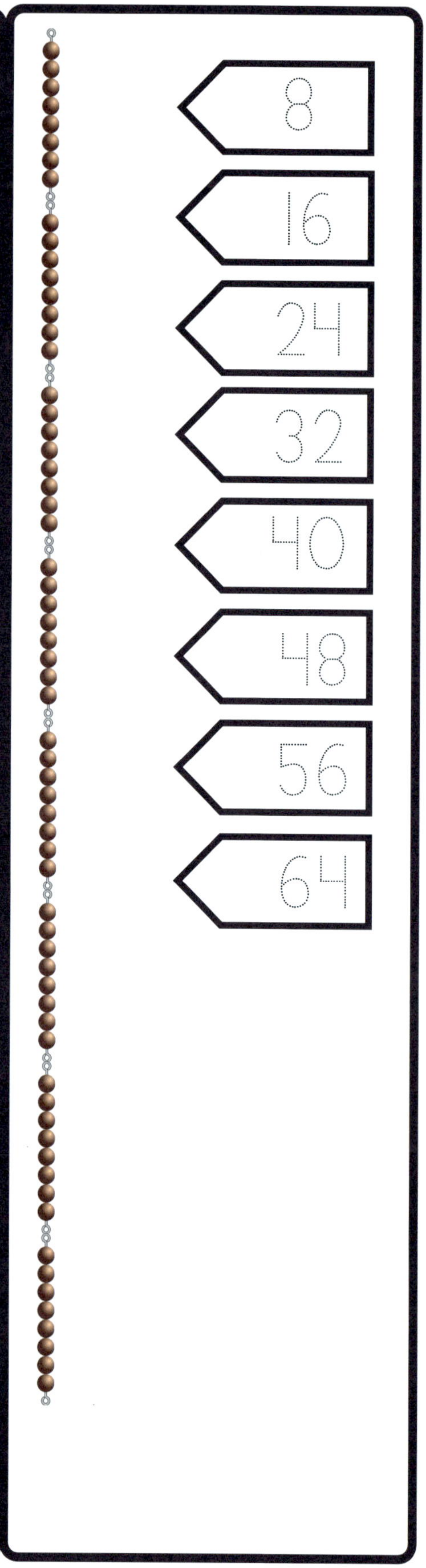
8
16
24
32
40
48
56
64

9
18
27
36
45
54
63
72
81

10
20
30
40
50
60
70
80
90
100

Color and Label Bead Chains

Directions:

1. Color the bead chain and the square their proper colors (**1=red**, **2=green**, **3=pink**, **4=yellow**, **5=turquoise**, **6=purple**, **7=white**, **8=brown**, **9=dark blue**).
2. Count the beads on the chain. When you get to an arrow, write the number inside the arrow. Children can use the "Bead Chains and Squares" worksheet for a control.
3. Count all the beads in the square. Write the number on the lines, filling in the statements. For children who are not reading, make sure to read all three statements to them.

*Note: I put the notation for the squaring equations on this worksheet just as a first exposure so they get a little familiar with it. The equations are abstract, and they do not need to know how to write them at this time. When you read the equation, it reads, "one squared equals one" after it is written.

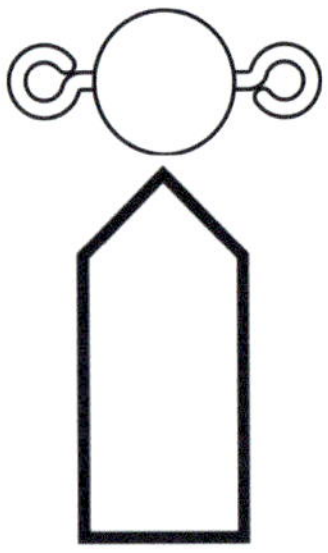

There is _________ bead on the 1-squared bead chain.

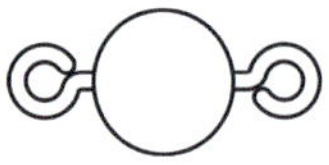

There is _______ bead in the 1-square.
The square of 1 is _________.

$$1^2 = \square$$

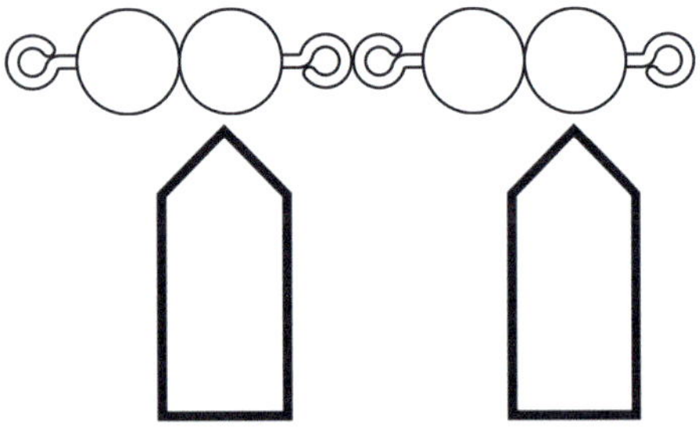

There are _________ beads on the 2-squared bead chain.

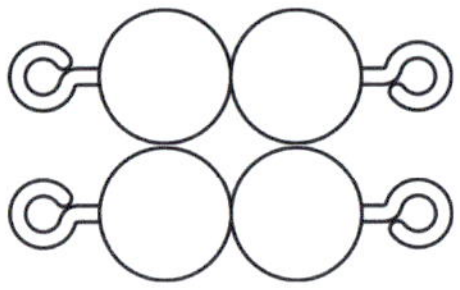

There are _______ beads in the 2-square.
The square of 2 is __________.

$$2^2 = \square$$

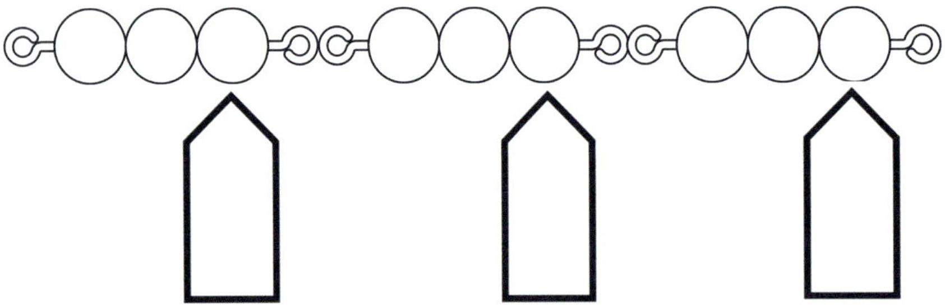

There are _________ beads on the 3-squared bead chain.

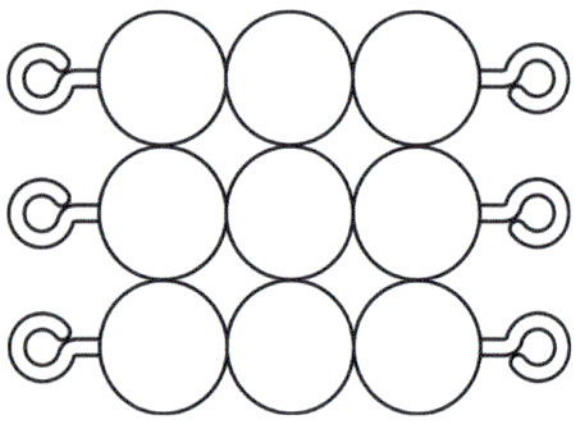

There are _______ beads in the 3-square.
The square of 3 is __________.

$$3^2 = \square$$

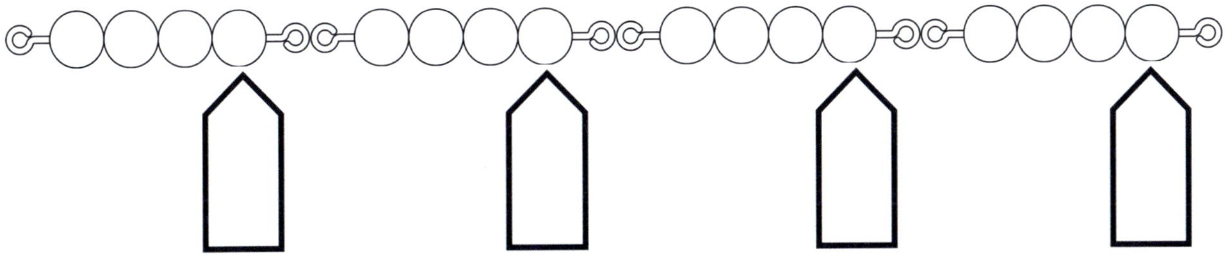

There are __________ beads on the 4-squared bead chain.

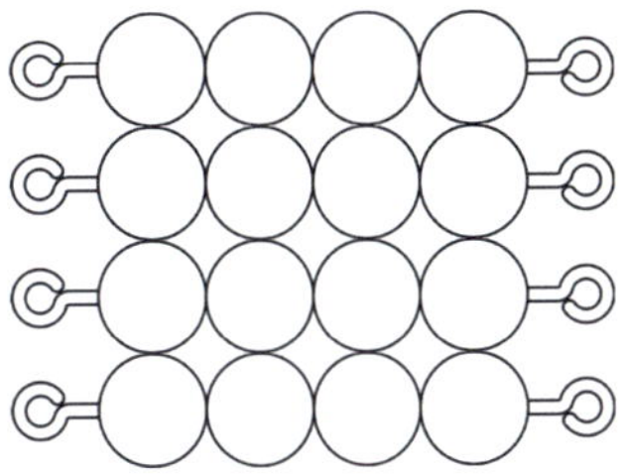

There are ________ beads in the 4-square.

The square of 4 is ___________.

$4^2 = \square$

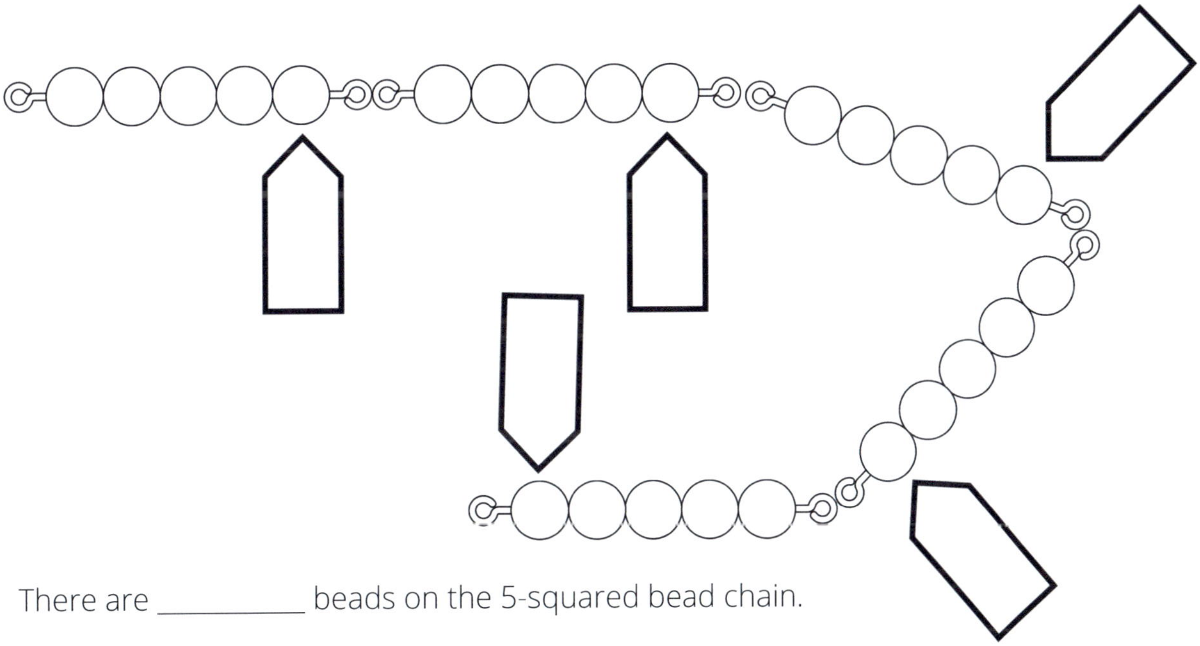

There are __________ beads on the 5-squared bead chain.

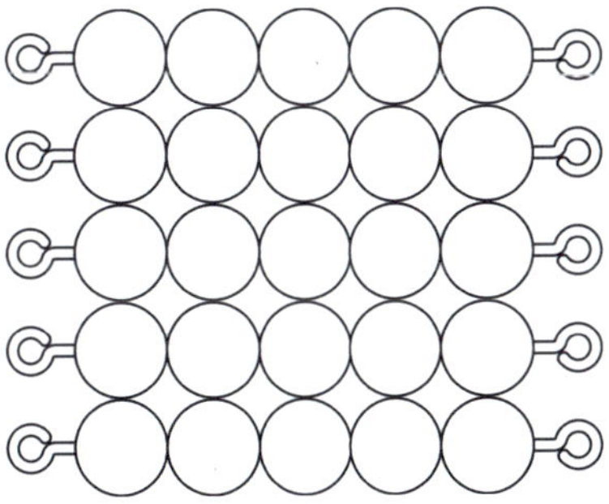

There are ________ beads in the 5-square.

The square of 5 is ___________.

$5^2 = \square$

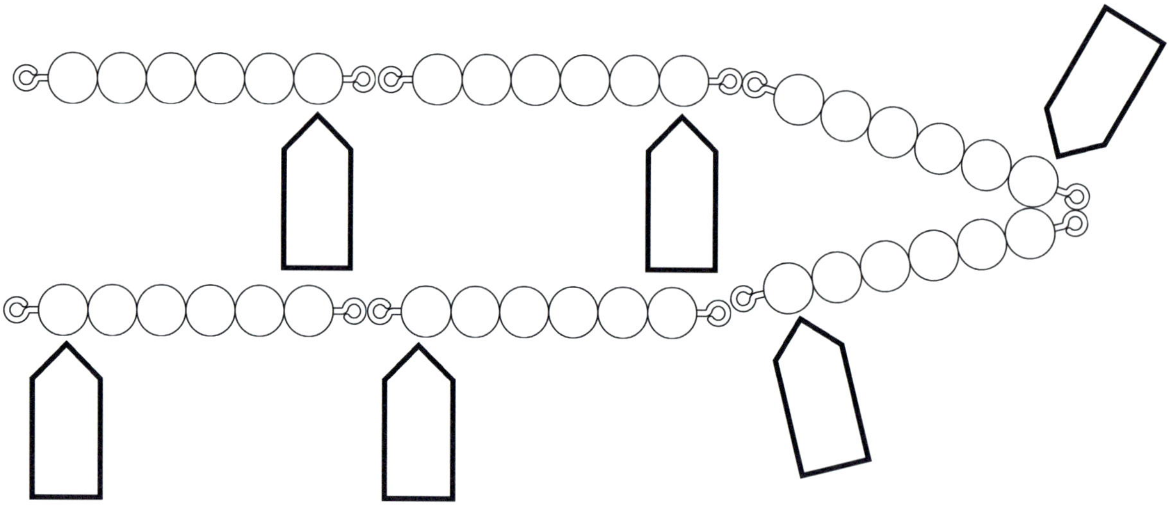

There are __________ beads on the 6-squared bead chain.

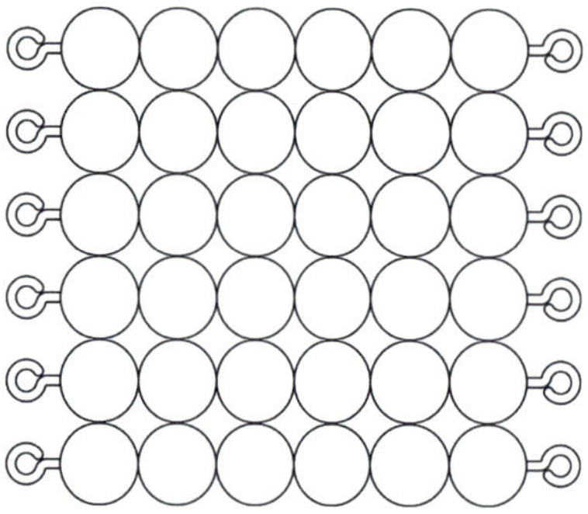

There are ________ beads in the 6-square.

The square of 6 is ____________.

$$6^2 = \square$$

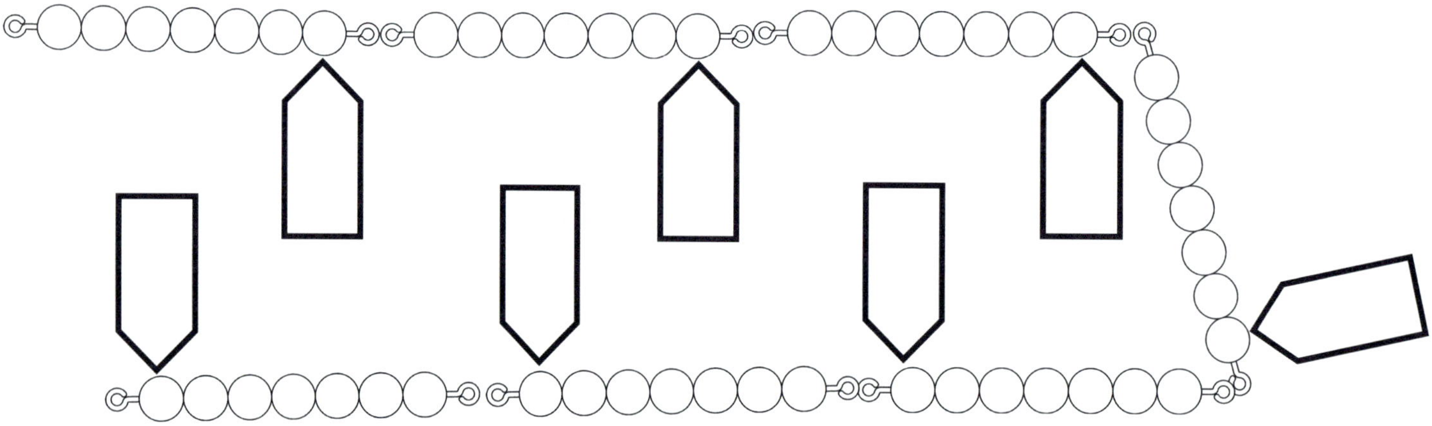

There are __________ beads on the 7-squared bead chain.

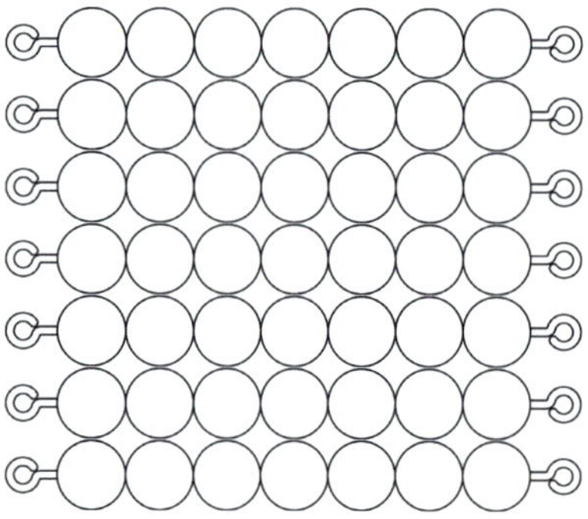

There are ________ beads in the 7-square.

The square of 7 is ____________.

$$7^2 = \square$$

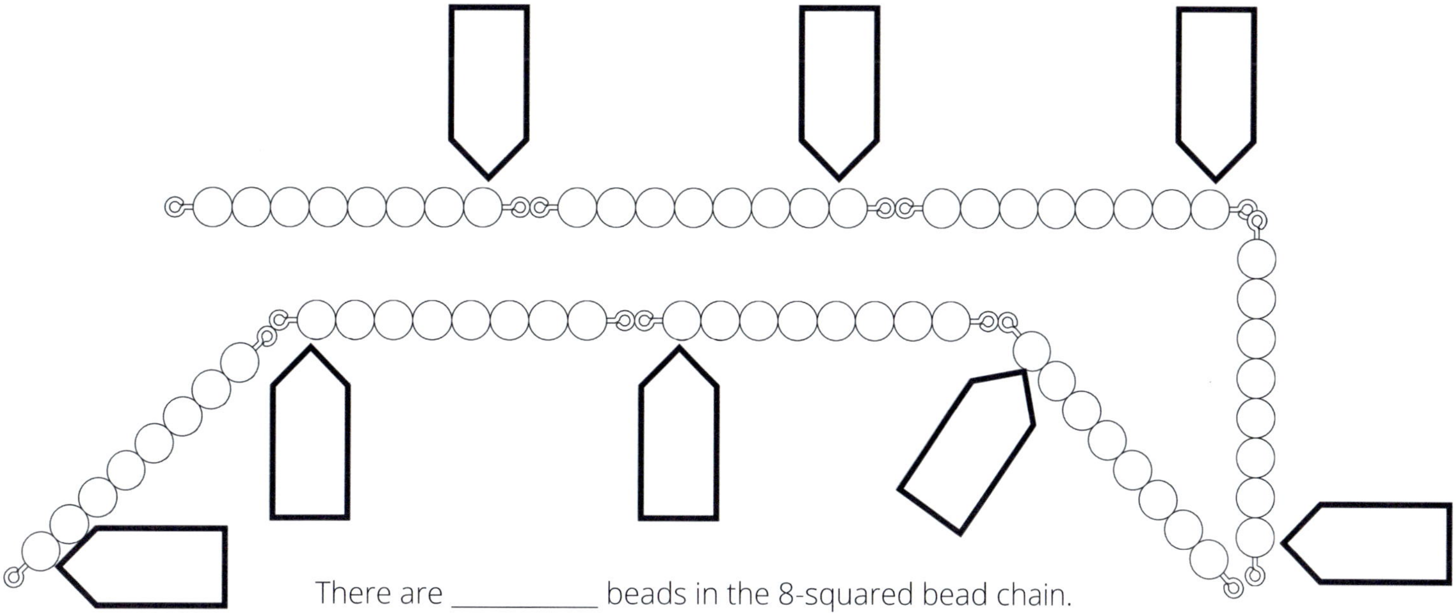

There are __________ beads in the 8-squared bead chain.

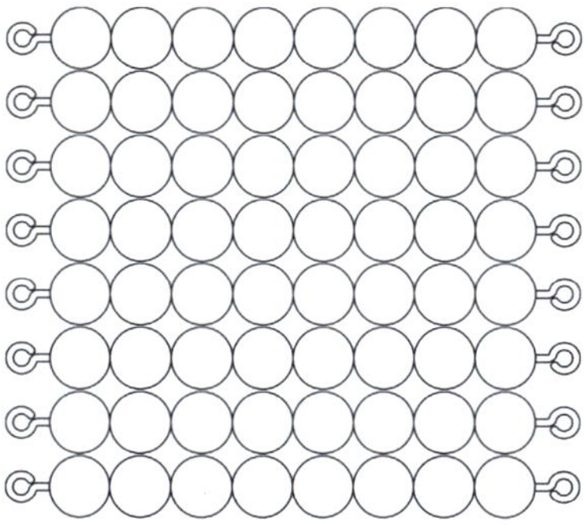

There are ________ beads on the 8-square.

The square of 8 is ____________.

$8^2 = \square$

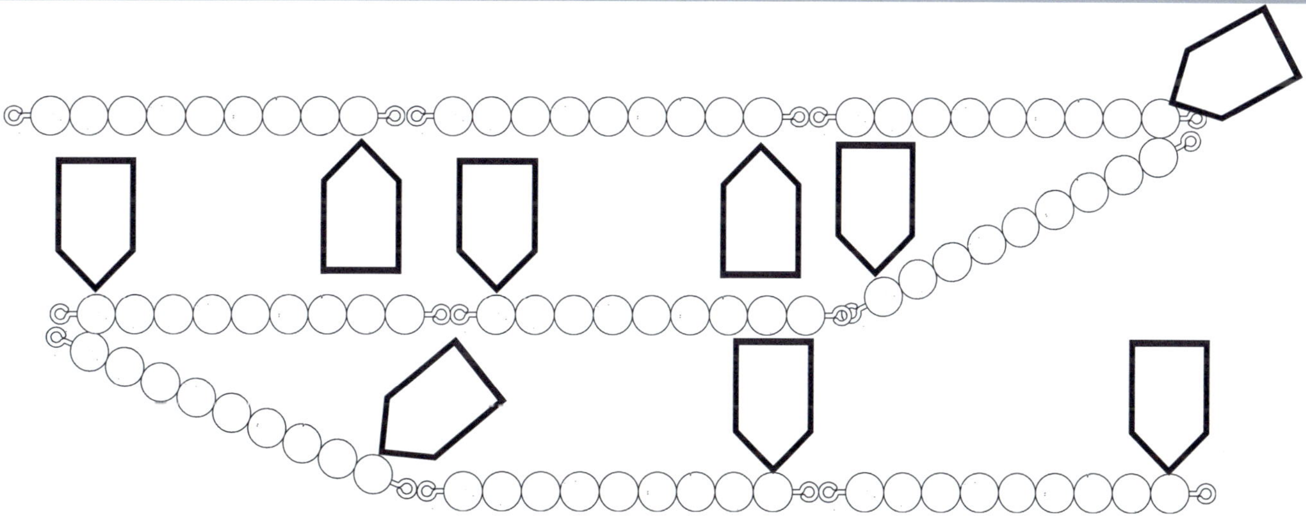

There are __________ beads on the 9-squared bead chain.

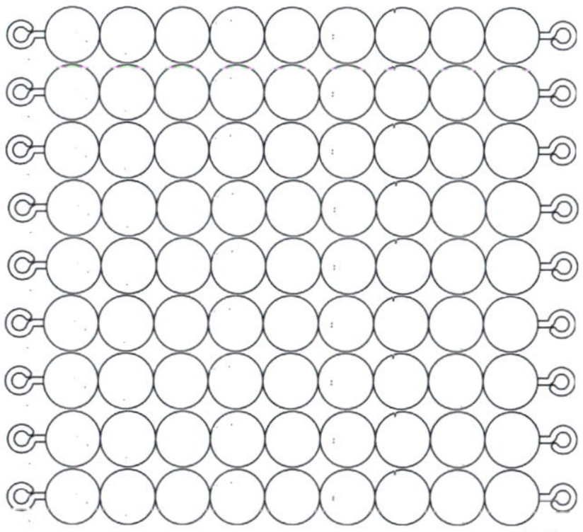

There are ________ beads in the 9-square.

The square of 9 is ____________.

$9^2 = \square$

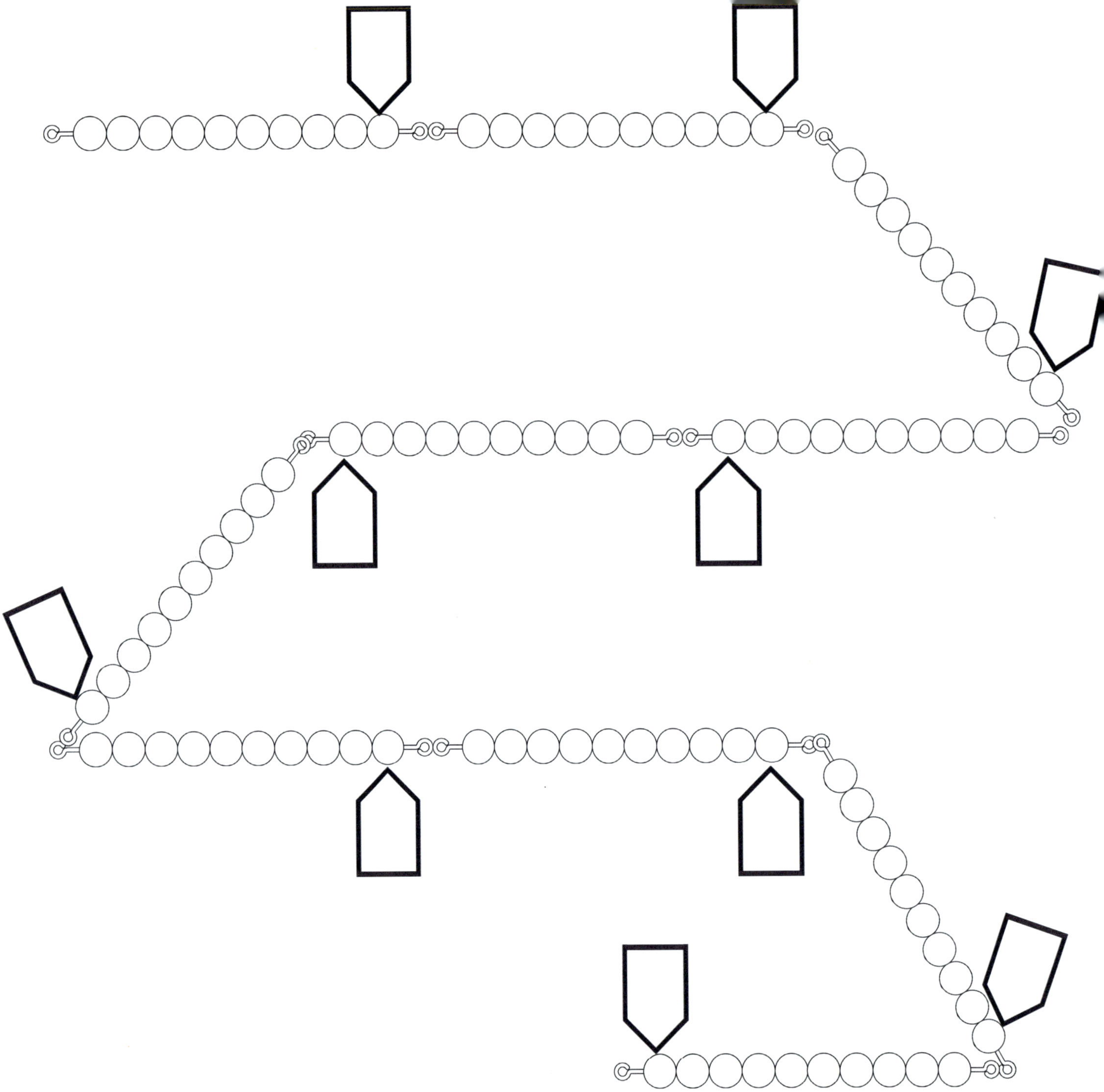

There are __________ beads on the 10-squared bead chain.

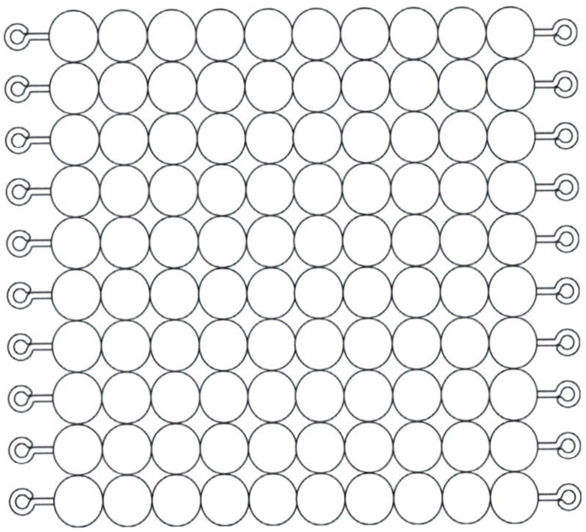

There are ________ beads in the 10-square.

The square of 10 is ____________.

$10^2 =$ ☐

1000 Chain

Directions: The 1000 Chain is a huge work that sparks the interest of the young child. You will need about 27 feet (8.2 m) to do this work!

1. Cut apart the groups of arrows for your child.
2. Cut apart the individual golden bead ten-bars. You will have 100 bars.
3. Cut apart the six pages of strips and tape them end to end. This will be your 1000 Chain mat.
4. Starting with the first ten-bar, glue or tape it down at the top of the 1000 Chain mat, leaving enough room to the right of the chain to fit the arrows. Continue for all 100 ten-bars.
5. Cut the arrows needed for the first ten-bar (green 1-9 and blue 10) and paste those into place on the 1000 Chain mat, to the right of the beads and pointing at the proper bead on the chain.
6. Using just one set of arrows at a time, invite your child paste the arrows into place on the 1000 Chain as you count the beads by tens.

NOTE: This work may take several sessions to complete. You can fold up the chain and place it in an envelope. Unfold and continue where your child left off.

Extensions:

A. Review counting by tens.

B. Ask your child to "find a number" - any number between 1-1000.

C. Count backwards from 1000-1.

1 2 3 4 5 6 7 8 9

210 220 230 240 250 260 270 280 290

10 20 30 40 50 60 70 80 90

310 320 330 340 350 360 370 380 390

110 120 130 140 150 160 170 180 190

410 420 430 440 450 460 470 480 490

510 520 530 540 550 560 570 580 590

810 820 830 840 850 860 870 880 890

610 620 630 640 650 660 670 680 690

910 920 930 940 950 960 970 980 990

710 720 730 740 750 760 770 780 790

100 200 300 400 500 600 700 800 900 1000

1000 Chain Beads - Page 1 of 2

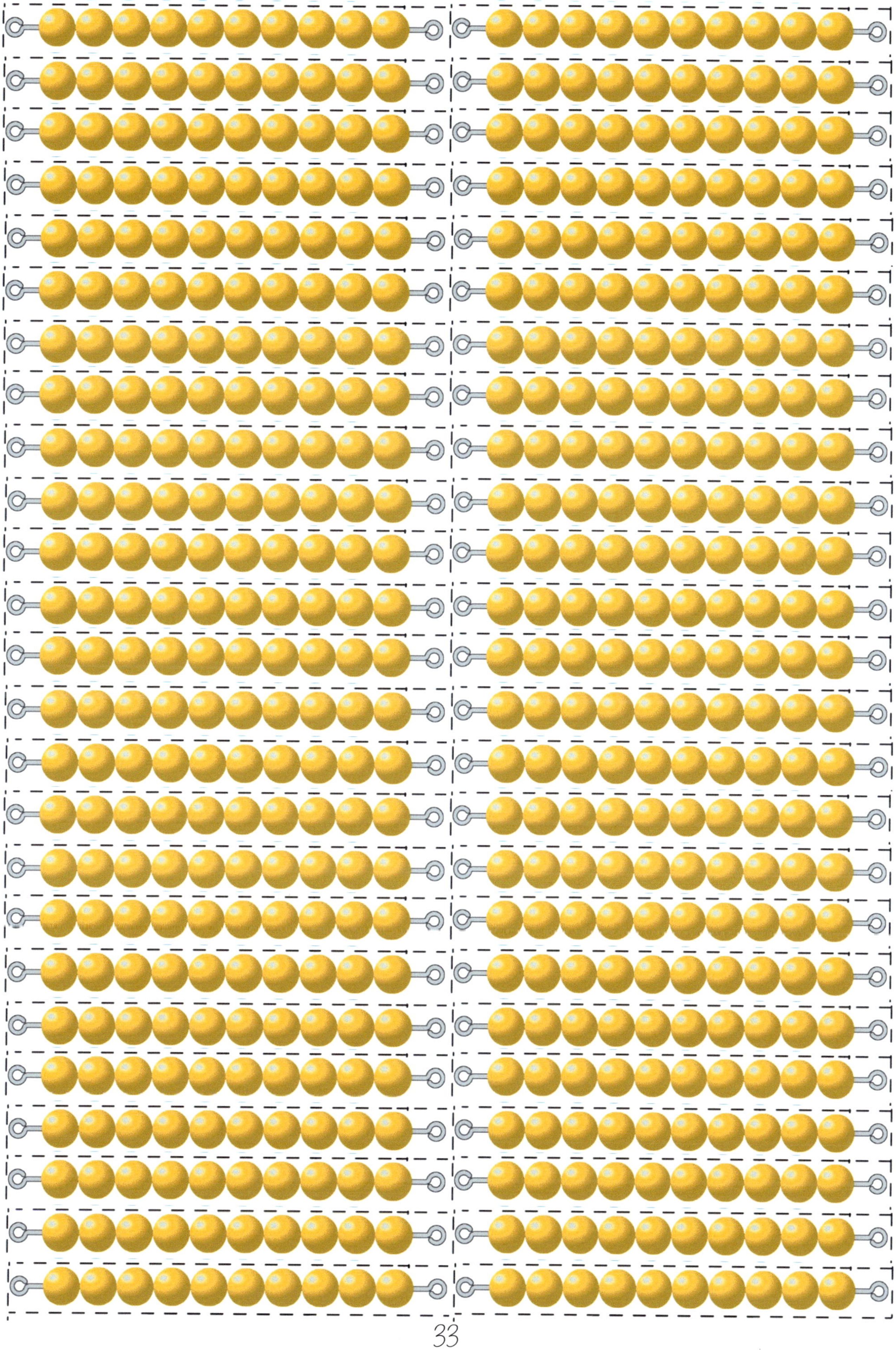

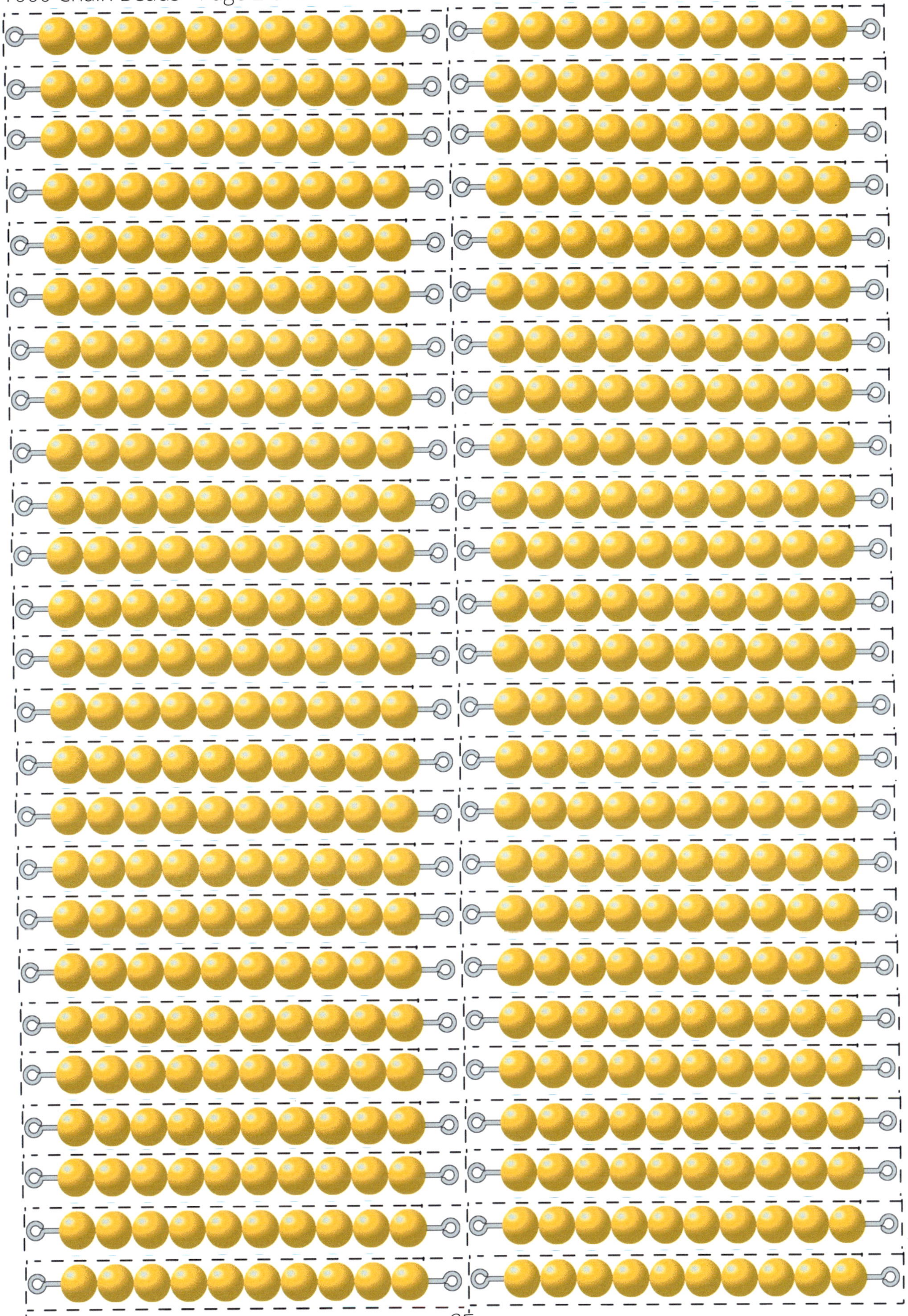

1000 Chain: Color and Count

This activity serves as a review for the 1000 Chain, as well as another opportunity for your child to practice counting to 1000 by tens.

Directions:

Invite your child to color each set of ten beads. When he/she gets to each arrow, write in the corresponding number. Your child can use the control chart below to mark off the numbers used OR use the chart as a cut-and-paste sheet.

NOTE:

1. If you use the chart as a cut-and-paste chart, you may want to only cut off one strip at a time.
2. This work may take several sessions to complete.
3. You also have the option of using the completed 1000 Chain activity as a control of error.

10	20	30	40	50	60	70	80	90	100
110	120	130	140	150	160	170	180	190	200
210	220	230	240	250	260	270	280	290	300
310	320	330	340	350	360	370	380	390	400
410	420	430	440	450	460	470	480	490	500
510	520	530	540	550	560	570	580	590	600
610	620	630	640	650	660	670	680	690	700
710	720	730	740	750	760	770	780	790	800
810	820	830	840	850	860	870	880	890	900
910	920	930	940	950	960	970	980	990	1000

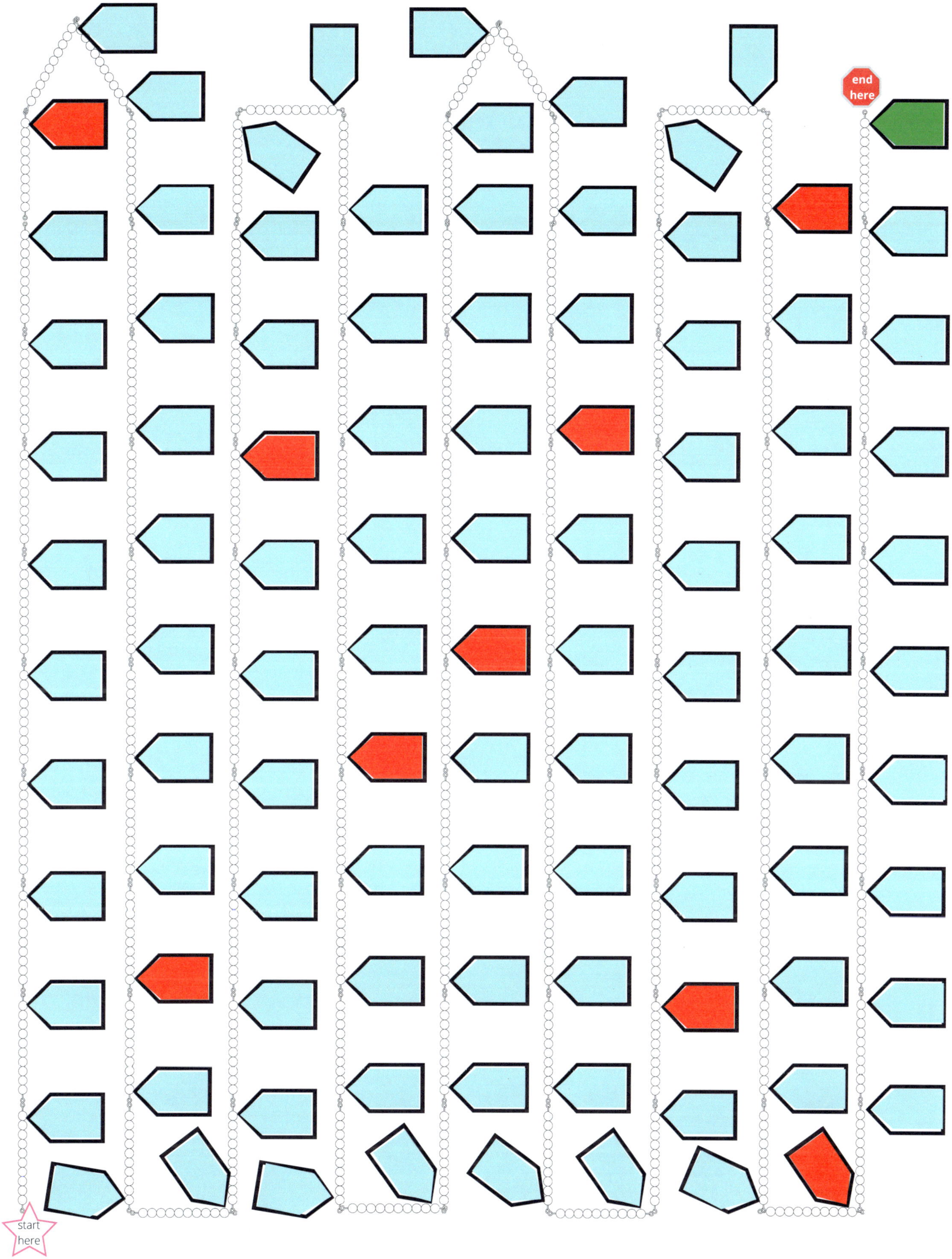
end here
start here

Making 1000 (and Full Layout - Options "A" and "B" for Hundred Squares)

Directions:

1. Show your child 1000 can be made more ways than just the linear chain of ten-bars. Show your child the 1000 chain he or she has already completed. Lay it out flat across the floor again. Say, "This is one thousand beads."
2. Then, show your child the hundred squares (see preparation instructions on the side of this page.) Stack the hundred squares and count 100-1000 by hundreds. Point to the stack of hundred squares and say, "This is one thousand beads." (I highly suggest making at least one set of ten hundred squares that are the thickness of a single bead.)
3. Lastly, show your child a one-thousand cube. (I suggest making at least **one** three-dimensional thousand cube to show your child using the printable in the workbook.) Say, "This is one thousand beads." Show your child the flat image of a one thousand cube. Say, "This is an image of one thousand beads. We sometimes use this picture to make it easier to show the beads on a flat surface, like this paper."

Directions for Making the Hundred Squares: Cut these out and glue them to an equal sized piece of cardboard. Ideally, the thickness of the square should match the thickness of one bead so that when you stack ten hundred squares, they will match the height of one thousand cube. You may need to stack two pieces of cardboard to reach the thickness of one bead. Make ten hundred squares for this activity. (Note: If you are going to do the "Full Layout" lesson, you will end up needing 45 of each unit.

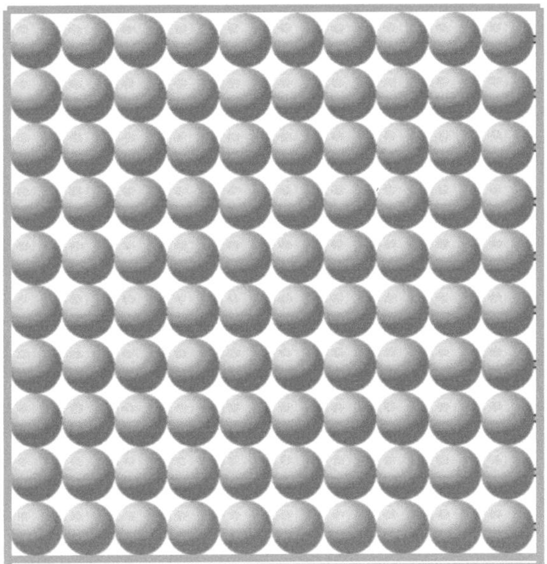
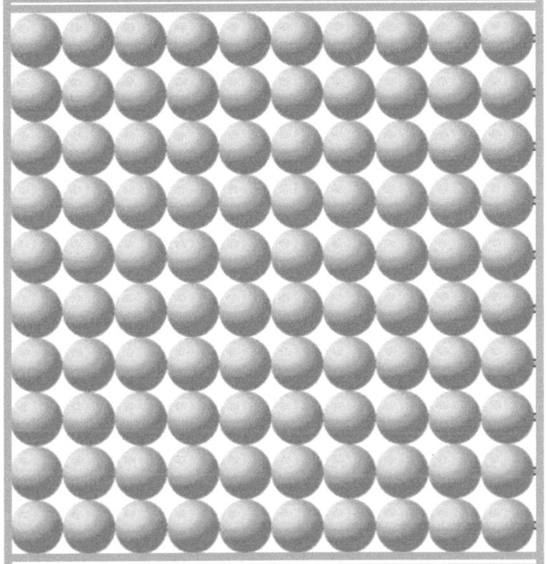
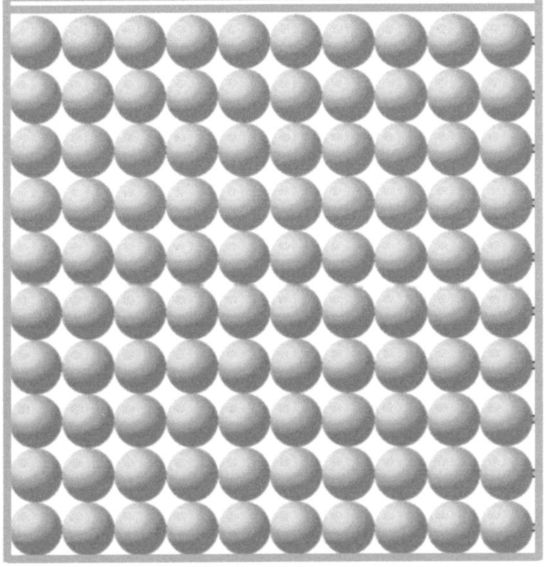

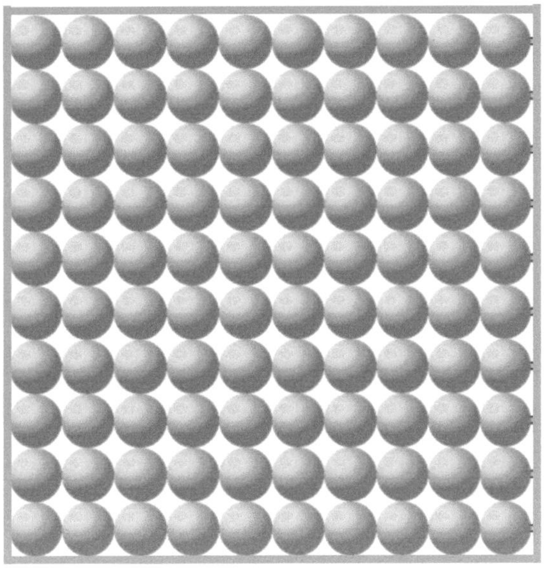

Options "A" and "B" for Hundred Squares

Option "B" for Thousand Cube

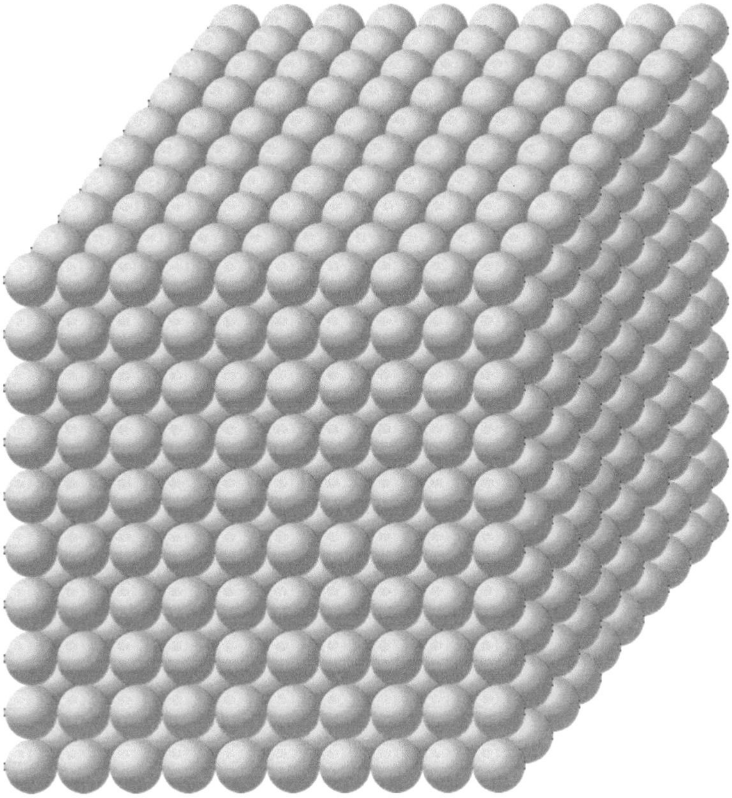

Two-Dimensional Thousand Cube: (Option B.) You may want to demonstrate how the two-dimensional thousand cube is a representation of the three-dimensional thousand cube.

Unit Beads Preparation: (Option A.) These are going to be used for many Golden Bead activities **if** you choose "Option A" for preparing your materials. Cut these out for individual beads. Optionally, you can add a dab of Hot Glue to the top of each one to make them three-dimensional and easier to handle.

3-D Thousand Cube - for "Making 1000" and Full Layout - "Option A"

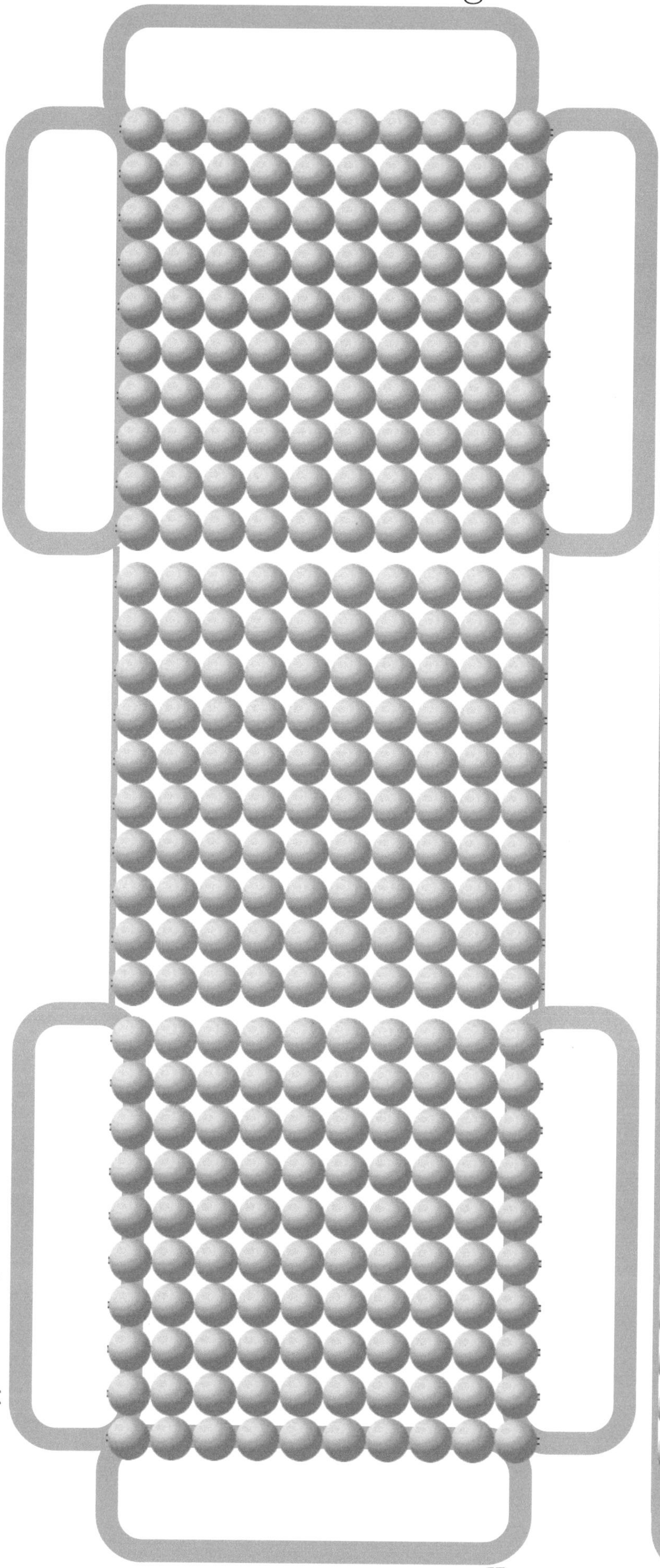

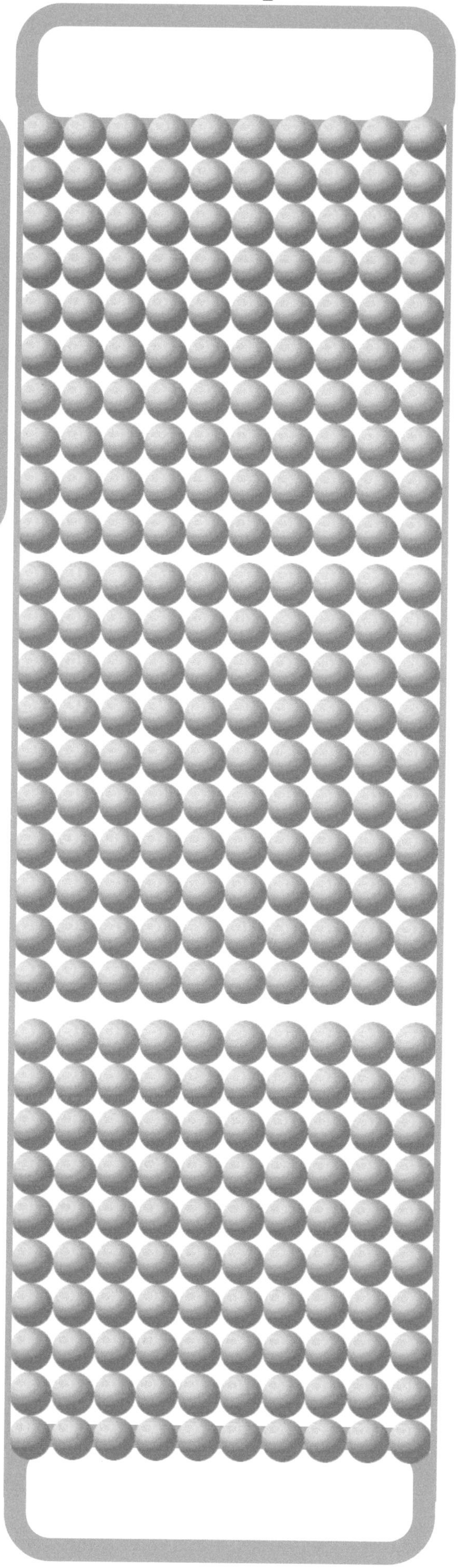

Directions:

1. Print (preferably on card stock, but regular paper will also work.) Cut out along the outside lines. Fold into a cube. Tape or glue folded flaps onto the inside of the cube (hot glue gun works best since it dries quickly and is sturdy). For the "Make 1000" lesson, you only need one cube. If you want to use these cubes for more Golden Beads lessons, you will need nine for the operations lessons and a total of 45 for the full layout. Read ahead to determine how many you should make.

Count and Compare 1000

Directions: Reinforce the tactile lesson "Making 1000" with this review worksheet. Invite your child to count by tens and hundreds to complete this lesson. Write the number in the space provided, then, using this sheet as a guide, answer the questions on the next page.

There are ______________ beads.

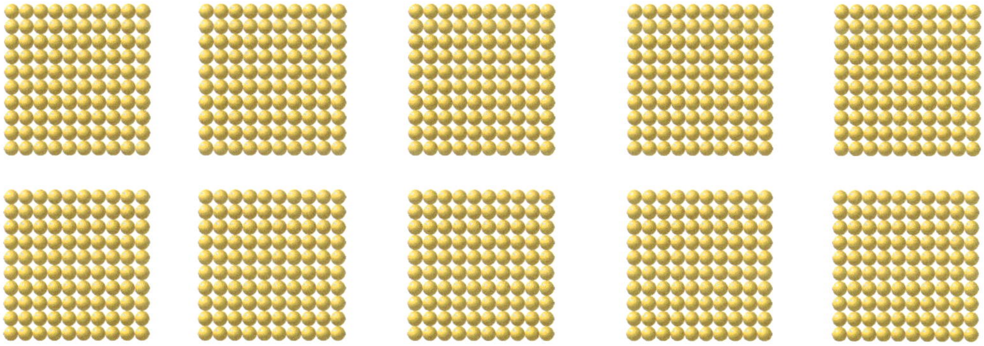

There are ______________ beads.

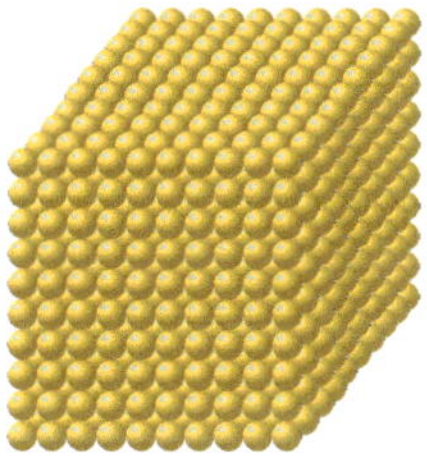

There are ______________ beads.

How many ten-bars are on the 1000 chain?

How many hundred-squares do you need to make 1000?

How many hundred-squares are in the thousand-cube?

How many unit beads are in the thousand-cube?

Make Your Own Thousand Cube

Directions: Cut out the hundred squares. Paste them on top of one another, but off center, covering the "star" with the top right corner of the next square so that each hundred square can be seen when all are pasted on the paper. You should be able to see part of the bead that is in the top right corner of each hundred square. This also gives it a three-dimensional look. As you paste, count by hundreds.

(This will help your child understand that even a flat, two-dimensional image of a one thousand cube contains one thousand units.)

Building Area:

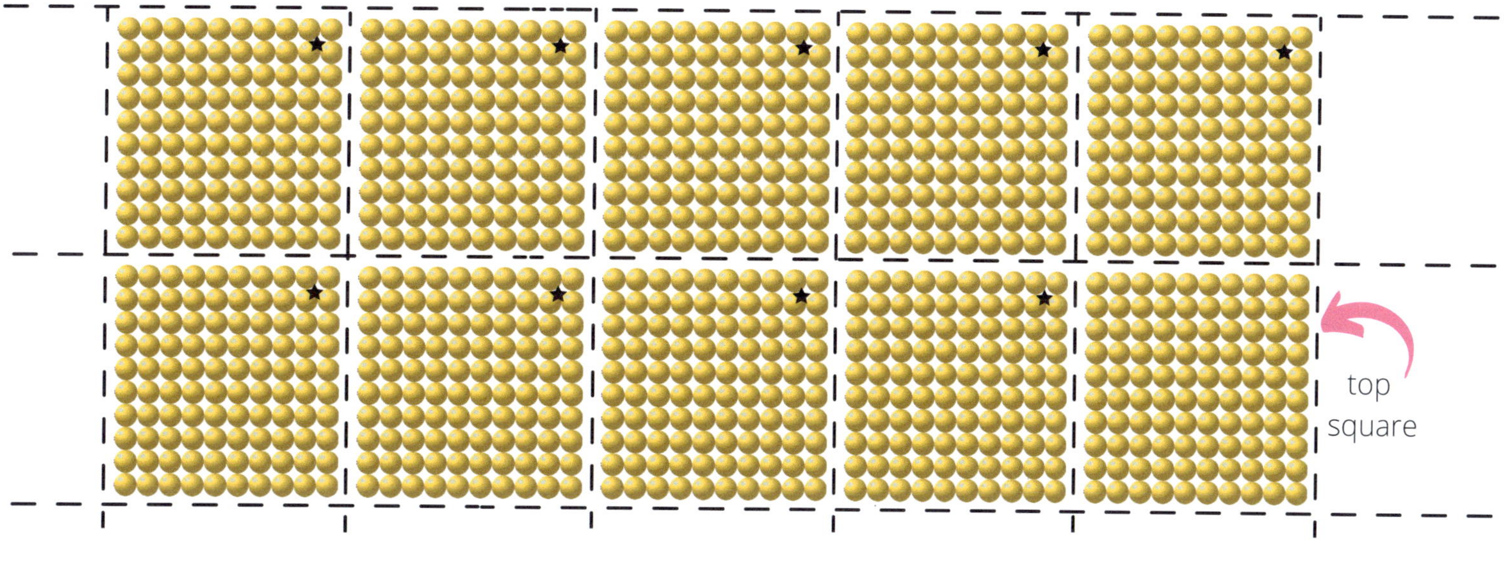

Introduction to the Decimal System

Directions: Use this lesson to reinforce the names and quantities of the materials.

1. Cut out the cards at the bottom of the page.
2. Pick up the unit card and say, "This is one unit." Place it to the far right section of the tray on the green "felt."
3. Pick up the ten card and say, "This is one ten." Place it on the blue "felt."
4. Pick up the hundred card and say, "This is one hundred." Place it on the red "felt."
5. Pick up the thousand card and say, "This is one thousand." Place it on the green "felt" at the far left section of the tray.
6. Glue these into place. This will serve as your control tray for the decimal system.

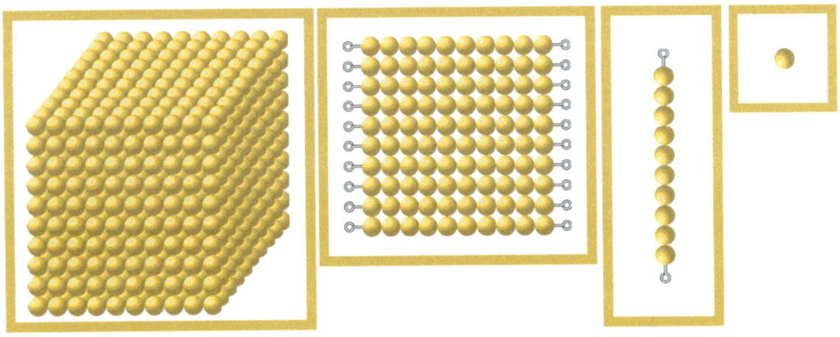

Decimal System: Second Period

Directions: Cut out the bead cards. Then, pointing to each spot in turn, ask for each of the cards in order. Mix them up. Then ask for each of the cards out of order. Lastly, mix up the cards and invite your child to place them in their proper spots.

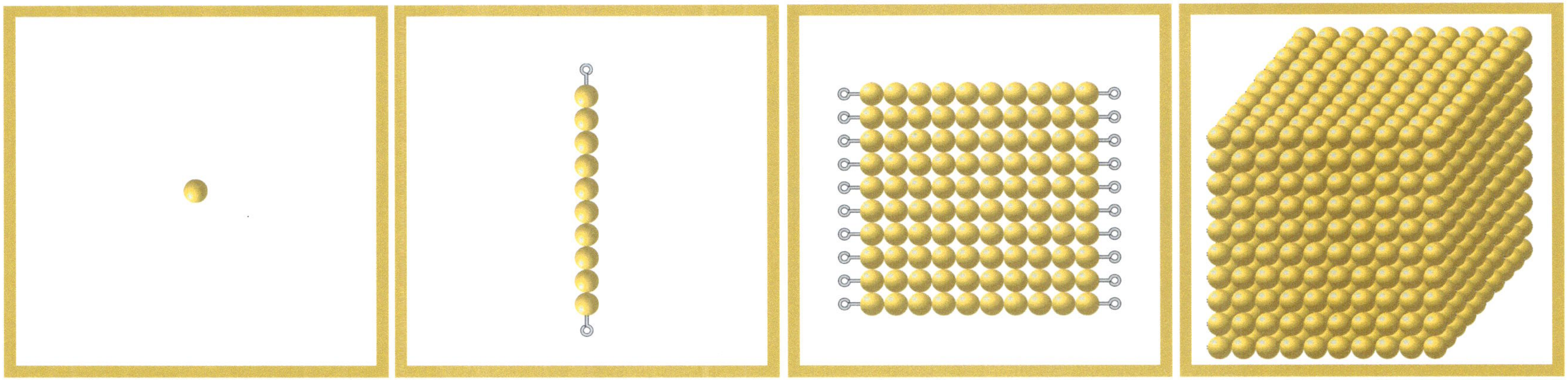

Golden Beads Match-Up

Directions: Draw a line from the bead to the name that represents the quantity. For non-readers, you can point to the words and read them. Trace words.

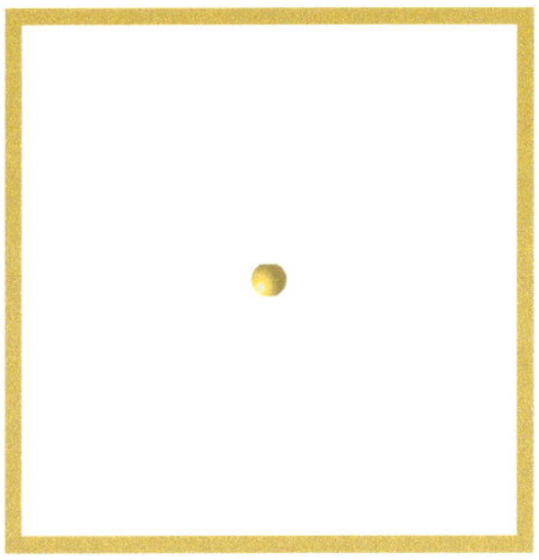

ten

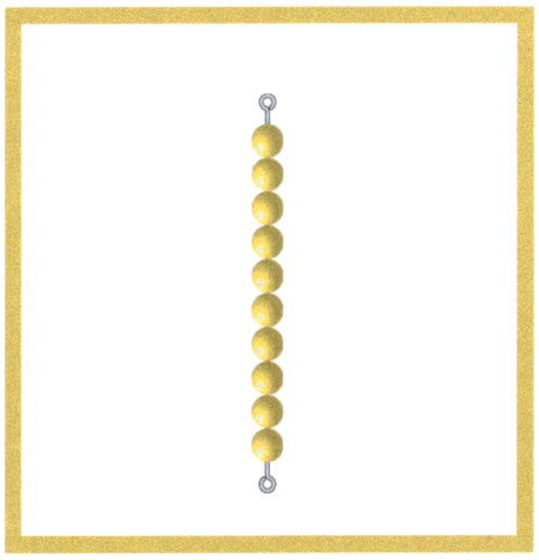

one thousand

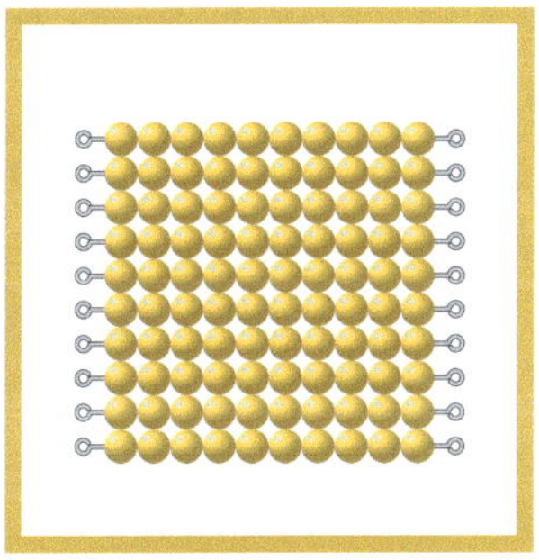

one hundred

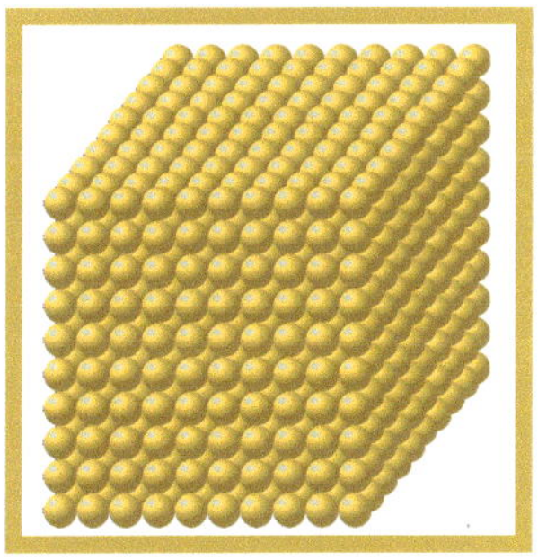

one

Golden Beads Match-Up

Directions: Draw a line from the bead to the symbol that represents the quantity. You will most likely need to read the numbers to your child, as we have not presented the decimal cards yet. This activity serves as a preview of the way we write numbers.
Trace numbers.

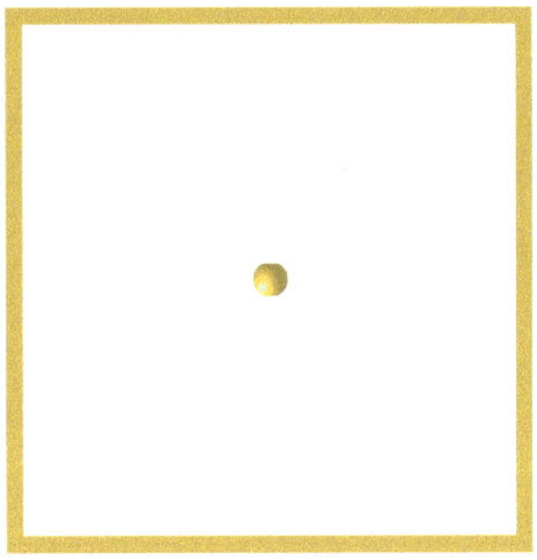

1000

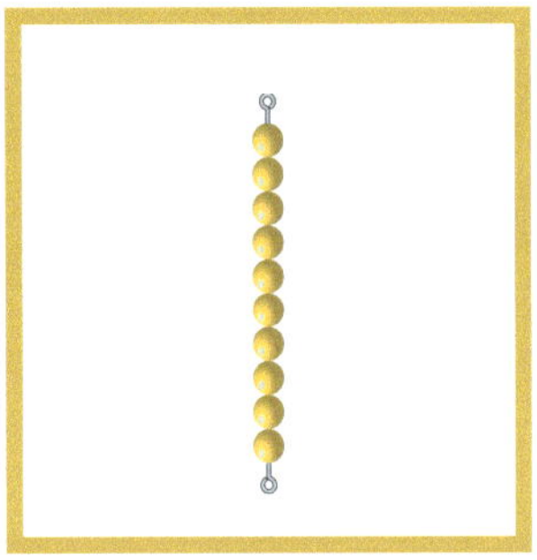

100

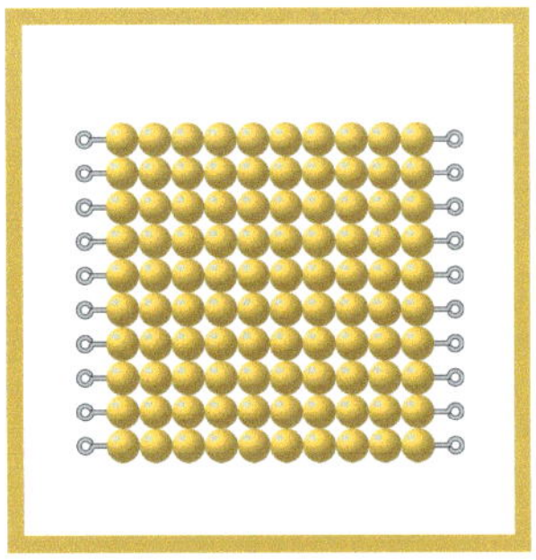

10

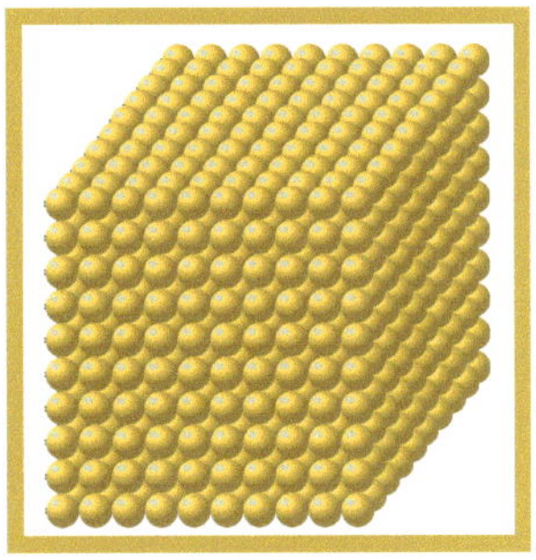

1

Decimal System - Complete Layout for Golden Beads

On this page, you will find several options for completing this work. First, choose an option. Then, follow the directions below the options. If you have a set of real Golden Beads, you may just need to use the printable versions to grow your collection to have 45 of each unit.

Option A: (See p. 49-53.) Large, three-dimensional cubes, hundred squares, ten-bars, and units. This is best for those who desire the closest replica of a physical Montessori lesson for the Golden beads without the actual materials. *These beads will be too large for the Operations Mats included in the workbook. You can choose to make your own mats or not use the mats at all. It is a personal preference.

Directions:

1. Print 45 three-dimensional cube pages on card stock paper.
2. Cut, fold, and hot glue or tape the cubes together.
3. Print and cut 45 hundred squares. (Regular paper is fine here.)
4. Cut equal sized cardboard squares for all of the hundred squares. The thickness of the cardboard should be equivalent to the width of a single bead. You may need to double up the cardboard layers to achieve the desired thickness. Paste the hundred squares onto the cardboard pieces.
5. Print and cut 45 ten-bars. (See p. 33 or cut apart hundred squares.)
6. Cut equal sized cardboard rectangles for all of the ten-bars. Paste the ten-bar cut outs onto the cardboard. (The thickness is not as crucial for the ten-bars.)
7. Print and cut 45 unit beads. You can put a dab of Hot Glue onto each bead before you cut them out to make them easier to handle and more like actual beads.

Option B: (See p. 49-53.) Large, two-dimensional cubes, hundred squares, ten-bars, and unit cards. Printables are included. You have permission to make copies for this purpose. Just print on 45 of each on card stock, cut, and you are ready to go!

Option C: Small, two-dimensional cards with pictures of the thousand cubes, hundred squares, ten-bars, and unit beads. Just cut and go! If you want, you can laminate them. All beads you need are included in the workbook, and **we will use this size for the Golden Beads Operations if you want to use the mats included in the workbook.**
(If you want to use the larger DIY Beads like Option A or Option B or real Golden Beads, you can make your own larger mats out of colored Construction Paper or Felt.)

Note: ***Whichever option you choose, keep the cards or DIY beads to use during many other Golden Bead activities in this workbook!***

Directions for Presenting: *see picture on next page.

1. Lay out a work mat or use a table large enough for the Option you chose. All beads will be laid out. If you are using **Option A**, it would be helpful to have nine small cups or containers to hold the unit beads. (Note: If you are using any of the cards, you can *lay them horizontally instead of stacking them.*)
2. Starting at the top right of the work area, place one unit. Say, "One unit."
3. Directly below the unit bead (vertically), place two units. Say, "One, two units."
4. Continue for all units. When you get to nine, say, "What comes after nine?" [Ten.]
5. Place one ten-bar at the top of the mat, to the left of the one unit bead. Say, "One ten."
6. Underneath the one ten-bar, place two ten-bars. Say, "One, two tens."
7. Continue for all tens. When you get to nine tens, say, "What comes after nine tens?" [Ten tens.] "Yes. Ten tens. Ten tens is one hundred."
8. Place the one hundred square to the left of the one ten-bar at the top of the mat. Say, "One hundred."
9. Underneath the one hundred, place two hundred squares, stacked on top of each other (like you are building a tower). Say, "One, two hundreds."
10. Continue for all of the hundreds, making stacks of the hundreds as you go. When you get to nine hundreds, ask, "What comes after nine hundreds?" [Ten hundreds.] "Yes, ten hundreds. Ten hundreds is one thousand."
11. Place the one thousand cube at the top of the mat to the left of the one hundred square. Say, "One thousand."
12. Underneath the one hundred, place two thousand cubes, stacked on top of each other (like you are building a tower) OR lay them side by side horizontally. Say, "One, two thousands."
13. Continue for all the thousands. (You will have a tall tower of nine thousand cubes when you get to the end if you chose **Option A**!)

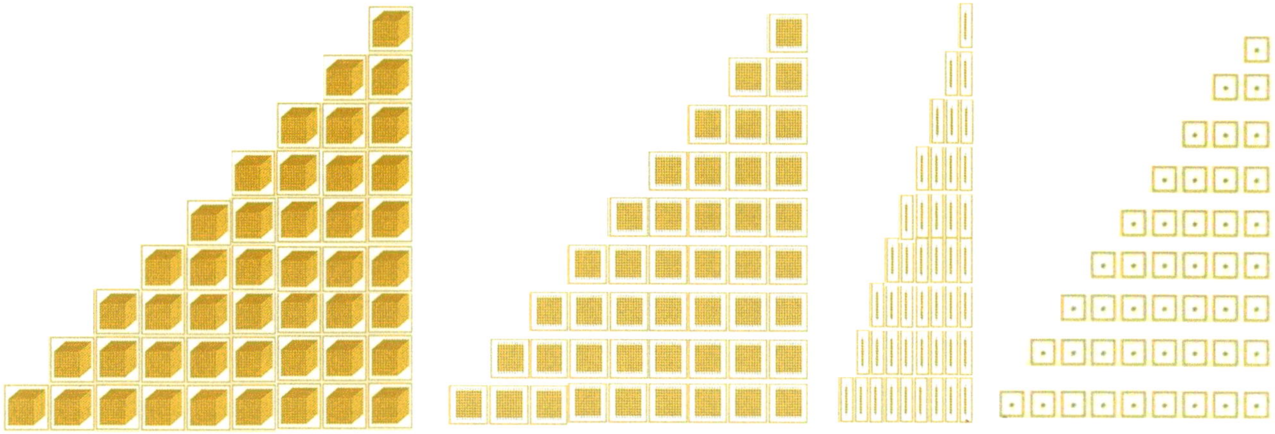

Golden Beads - You will need 45 of each type. Laminate and keep them safe to reuse.

Golden Beads - You will need 45 of each type. Laminate and keep them safe to reuse. There are 3 extra of each type.

Golden Beads Bank

*This is for the smaller set of Golden Beads (Option C) that fit on the Operations Mats.

Remove this mat from the workbook, and use it when you are working with the Golden Beads to keep them organized.

When you are not using them, the Golden Beads can be stored in the envelope you can make by folding up the bottom section of this page.

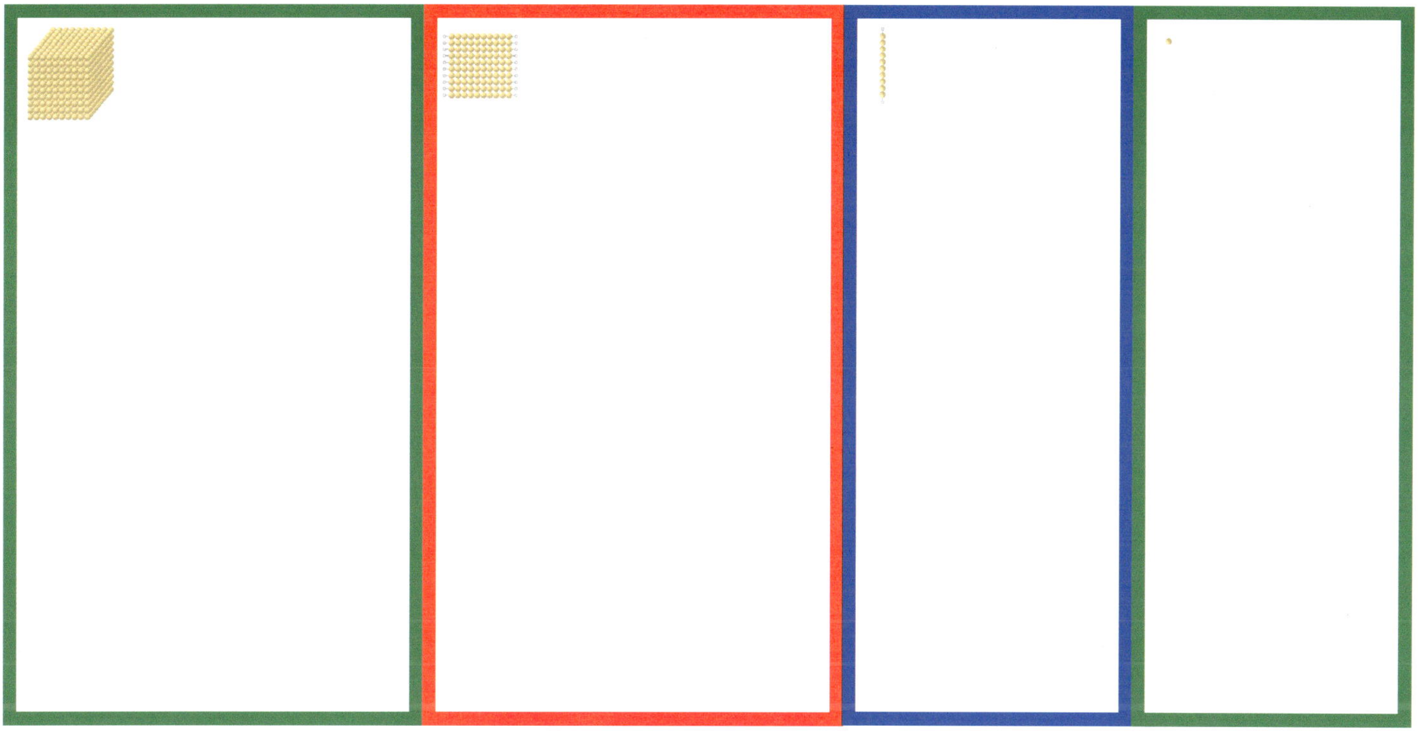

Fold this side up. Then glue or staple the sides and dividers.

Golden Beads
Storage
"Envelope"

Which One is Missing?

Directions: Invite your child to tell you which unit is missing! First, review the quantity of materials below. Then, invite your child to tell you which one is missing. Circle the name of the quantity that is missing.

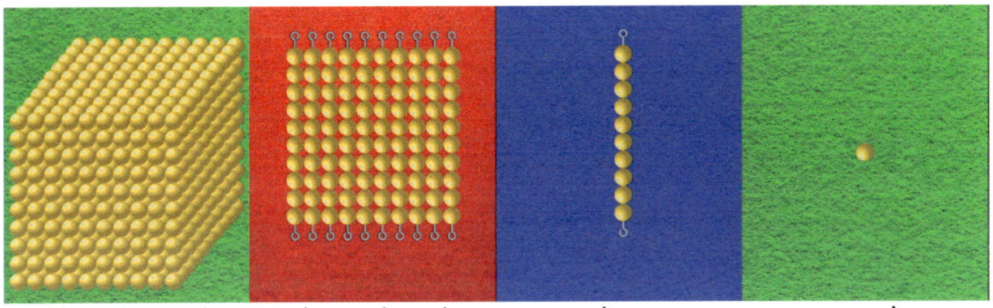

thousand hundred ten unit

unit
ten
hundred
thousand

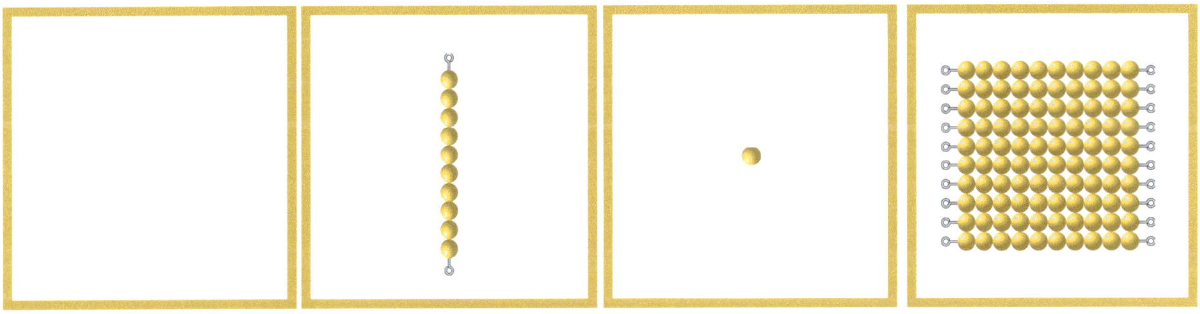

unit
ten
hundred
thousand

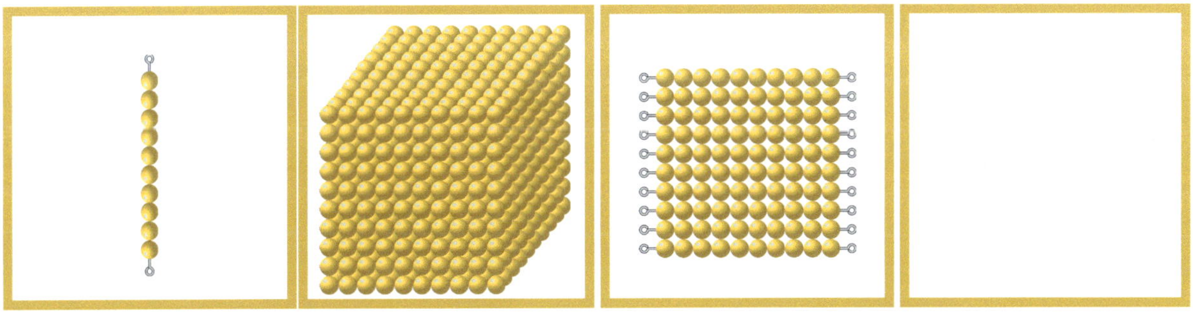

unit
ten
hundred
thousand

unit
ten
hundred
thousand

Relationship of Beads - 1

Match each unit bead to a corresponding bead on the ten-bead bar by drawing a line from the unit bead to a single bead on the ten-bead bar. Then count the number of unit beads in each column.

Units:

Units:

Relationship of Beads - 2

Match each ten-bead bar to a row of ten beads on the hundred square by drawing a line from the ten-bead bar to the hook that starts the row of ten beads on the hundred square. Then count the number of tens in each column.

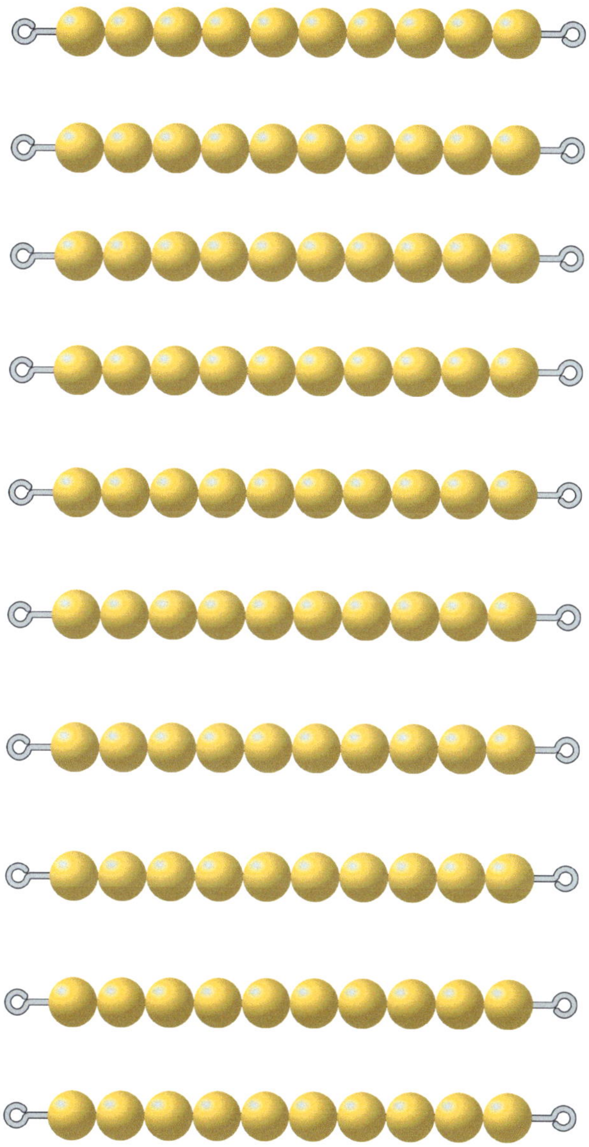

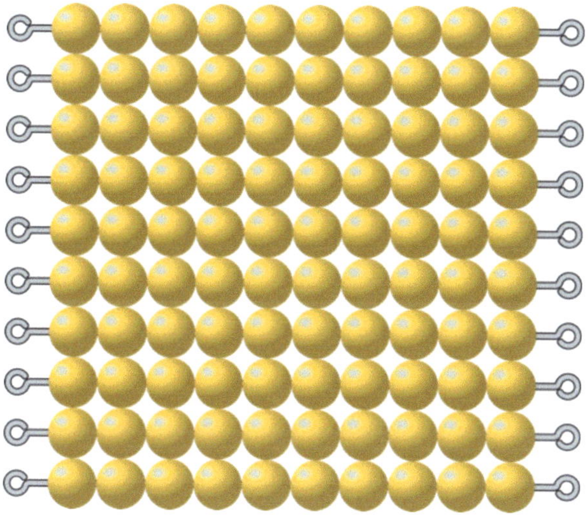

Tens:

Tens:

Relationship of Beads - 3

Match each hundred square to an arrow that points to a hundred square within the thousand cube. (Note: The thousand cube is enlarged, so if this is confusing for your child, please skip this activity for now.)

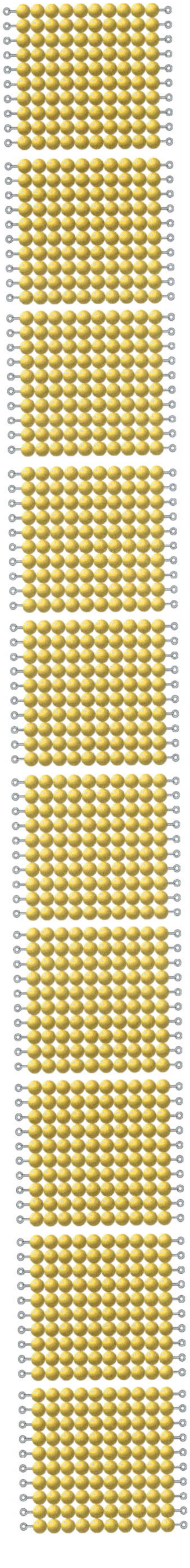

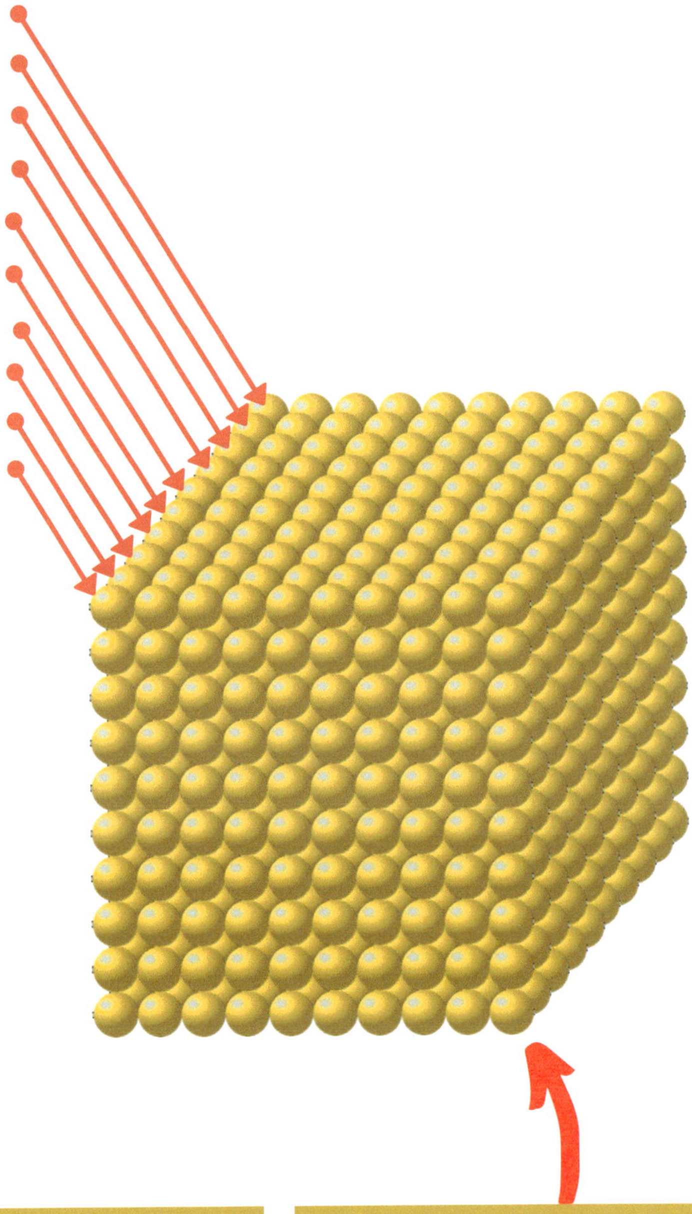

Hundreds:

Hundreds:

"Give Me an Amount" Golden Beads Bank

Directions: Cut these out and sort them by category to use in the next activity. You will have leftover bead cards.

Give Me an Amount

Using your "bank" of Golden Bead cards, invite your child to paste the amount of beads asked for in each box. This will get progressively harder.

Please give me three tens.

Please give me five hundreds..

Please give me two thousands..

Please give me six units..

Give Me an Amount

Using your "bank" of Golden Bead cards, invite your child to paste the amount of beads asked for in each box. This will get progressively harder. The bead cards can overlap! You may also want to cut out the tens and units so that there is more room for pasting.

Please give me two tens and four units.

Please give me six hundreds and one ten.

Please give me three thousands and four hundreds.

Please give me one thousand and six units.

Give Me an Amount

Using your "bank" of Golden Bead cards, invite your child to paste the amount of beads asked for in each box. This will get progressively harder. The bead cards can overlap! You may also want to cut out the tens and units so that there is more room for pasting.

Please give me two thousands, one hundred, and three units.

Please give me one thousand, four hundreds, and one ten.

Please give me three thousands, nine hundreds, and one unit.

Please give me one thousand, one hundred, seven tens, and five units.

Introduction to the Decimal System Cards

This is how we write one unit. [Point.]
How many zeros are in one unit? [None.]

1

This is how we write one ten. [Point.]
How many zeros are in one ten? [One.]

10

This is how we write one hundred. [Point.]
How many zeros are in one hundred? [Two.]

100

This is how we write one thousand. [Point.]
How many zeros are in one thousand? [Three.]

1000

Second and Third Period for Introduction to Decimal System:

Second Period: Show me 10. [ ] Show me 1. [ ] Show me 1000. [ ] Show me 100. [ ]

Third Period: What is this? (Point to each number, out of order.) 1 [ ] 10 [ ] 100 [ ] 1 [ ]

*use the brackets to check off what your child has mastered.

Decimal System Cards - Full Layout

Introduction: We will introduce the Decimal System Cards and make sure your child can "read" these cards before moving on to using them for forming numbers and performing operations. There is one set of large Decimal Cards and three sets of Small Decimal Cards. The Large Decimal Cards will be used for the preliminary exercises.

Preparation: Cut and **laminate** the cards. It is important to laminate these cards if you can. You can use clear packing tape or contact paper to laminate them if you do not have a laminator.

Present: Using work rug on the floor or the table surface, lay out the cards as outlined below:

1. Start with the units. Place the "1" card at the top right of your work surface. (Make sure there is enough room to lay out all 9 units.) Say, "One unit."
2. Place the "2" underneath the "1" and say, "Two units." Continue for all 9 units.
3. Ask, "What comes after nine?" ["Yes, one ten!"]
4. Place the "10" to the left of the "1" at the top row. Count by tens now, like this, "One ten. (Place 10 card.) Two tens. (Place 20 underneath.) Three tens. (Place 30 underneath.)" And so on until 90.
5. Ask, "What comes after nine tens?" ["Yes, ten tens! Which is one hundred."]
6. Place the "100" to the left of the "10" in the top row.
7. Continue with the hundreds, making sure to name them, "two hundreds, three hundreds, four hundreds, etc."
8. Ask, "What comes after nine hundreds?" ["Yes, ten hundreds! Which is one thousand."]
9. Place the "1000" to the left of the "100" in the top row.
10. Continue with the thousands, making sure to name them, "two thousands, three thousands," etc. Stop at 9000.

1000	100	10	1
2000	200	20	2
3000	300	30	3
4000	400	40	4
5000	500	50	5
6000	600	60	6
7000	700	70	7
8000	800	80	8
9000	900	90	9

Decimal System Cards - Fetch Game

Introduction: This is an interactive activity! Use the large Decimal Cards that you cut and laminated for the "Full Layout" activity.

Directions:

1. Invite your child to lay out all of the decimal system cards on the work surface (table or work rug) in the same manner as the original presentation. You may help as needed.
2. Go across the room and call out commands from the table below. Make notes as you see fit in the table. A simple check-mark for correctness works well for record keeping.
3. If your child brings you the wrong card, say, "I asked for two hundreds. This is two thousands. Can you bring me two hundreds? Two hundreds has two zeros." Give your child many opportunities to be right!

☐	1	Please bring me four units.
☐	2	Please bring me six units.
☐	3	Please bring me one unit.
☐	4	Please bring me seven units.
☐	5	Please bring me five tens.
☐	6	Please bring me two tens.
☐	7	Please bring me nine tens.
☐	8	Please bring me three tens.
☐	9	Please bring me one hundred.
☐	10	Please bring me eight hundreds.
☐	11	Please bring me seven hundreds.
☐	12	Please bring me four hundreds.
☐	13	Please bring me three thousands.
☐	14	Please bring me five thousands.
☐	15	Please bring me nine thousands.
☐	16	Please bring me two units and three units.
☐	17	Please bring me one ten and five tens.
☐	18	Please bring me seven hundreds and four hundreds.
☐	19	Please bring me six thousands and eight thousands.
☐	20	Please bring me three units and two tens.
☐	21	Please bring me two units and one hundred.
☐	22	Please bring me three tens, four hundreds, and one thousand.
☐	23	Please bring me six units and seven thousands.
☐	24	Please bring me two tens, six hundreds, and three thousands.
☐	25	Please bring me one unit, three tens, five hundreds, and eight thousands.
☐	26	Please bring me five tens and six thousands.
☐	27	Please bring me nine units and four hundreds.
☐	28	Please bring me three tens and two hundreds.
☐	29	Please bring me eight units and seven tens.
☐	30	Please bring me four units, two tens, and three thousands.

place the card here in the "inspection zone"

Decimal System Cards

1	2	3	4	5
6	7	8	9	90

10	20	100
30	40	200
50	60	300
70	80	400

500	5000
600	4000
700	2000
800	3000
900	1000

6000

7000

8000

9000

1000	100	10	1
2000	200	20	2
3000	300	30	3
4000	400	40	4
5000	500	50	5
6000	600	60	6
7000	700	70	7
8000	800	80	8
9000	900	90	9

1000	100	10	1
2000	200	20	2
3000	300	30	3
4000	400	40	4
5000	500	50	5
6000	600	60	6
7000	700	70	7
8000	800	80	8
9000	900	90	9

1000	100	10	1
2000	200	20	2
3000	300	30	3
4000	400	40	4
5000	500	50	5
6000	600	60	6
7000	700	70	7
8000	800	80	8
9000	900	90	9

Circle the Amount

Directions: First, review the decimal system cards for 1000, 100, 10, and 1. Next, if your child is not yet reading, read the directions to your child in each box. Invite your child to circle the correct value.

1000	100	10	1
one thousand	one hundred	one ten	one unit

Circle four tens.

4000
400
40
4

Circle five units.

5000
500
50
5

Circle nine thousands.

9000
900
90
9

Circle two hundreds.

2000
200
20
2

Circle three units.

3000

300

30

3

Circle eight tens

8000

800

80

8

Circle six tens.

6000

600

60

6

Circle seven thousands.

7000

700

70

7

Circle one hundred.

1000

100

10

1

Circle two units.

2000

200

20

2

Follow the Directions Game

1. Find one unit. Put your finger on one unit. Shade in one unit with green.
2. Find three tens. Put your finger on three tens. Draw a box around three tens.
3. Find five thousands. Put your finger on five thousands. Draw spikes coming out of five thousands.
4. Find seven hundreds. Put your finger on seven hundreds. Circle seven hundreds.
5. Find nine units. Put your finger on nine units. Color nine units any way you want.
6. Find six thousands. Put your finger on six thousands. How many zeros does six thousands have? Draw an "X" on six thousands.
7. Find eight tens. Put your finger on eight tens. How many zeros does eight tens have? Shade in the inside of the zero in the eight tens with green.
8. Find two hundreds. Put your finger on two hundreds. How many zeros does two hundreds have? Draw a smiley face inside the zeros of the two hundreds.
9. Find four units. Put your finger on the four units. How many zeros does four units have? Trace the number for four units.

5000

80

1

700

200

6000

9

30

4

Quantity and Symbol

Point to each category in turn. Say, "This is one thousand beads. (point to cube)." Then point to the 1000 number card and say, "This says one thousand." Repeat for 100, 10, and 1. Then follow up with the second period lesson, "Show me," for each number and bead amount. Lastly, try the third period lesson, "What is this?" for each number and quantity.

1000:

1. Present. [.] Number. [.] Quantity
2. Show me. [.] Number. [.] Quantity
3. What is this? [.] Number. [.] Quantity

100:

1. Present. [.] Number. [.] Quantity
2. Show me. [.] Number. [.] Quantity
3. What is this? [.] Number. [.] Quantity

10:

1. Present. [.] Number. [.] Quantity
2. Show me. [.] Number. [.] Quantity
3. What is this? [.] Number. [.] Quantity

1:

1. Present. [.] Number. [.] Quantity
2. Show me. [.] Number. [.] Quantity
3. What is this? [.] Number. [.] Quantity

Bird's Eye View Complete Layout of Quantity and Symbol

Materials: You should still have the materials from the "Decimal System - Complete Layout for Golden Beads" and the "Decimal System Cards - Complete Layout." Use these materials. Any of the "Options" you chose previously for the Golden Beads will work for this activity.

- 45 thousand cubes
- 45 hundred squares
- 45 ten-bars
- 45 unit beads

Directions: Present this material similarly to how you did for the Beads by themselves and the Cards by themselves. First, invite your child to lay out *all* of the decimal cards, leaving room to stack or place the beads to the left of the decimal cards. Next, invite your child to count out all of the beads and put them in their proper places. This activity could take 45 minutes or longer for your child to complete! You may want to set this up somewhere that you can leave the activity set out for your child to complete it in stages or over multiple days.

Note: For 3-dimensional beads, stacking them like a tower is the best way to go. If you have 2-dimensional beads (like the printed bead cards), lay them horizontally to the left of the number cards. Your columns will be spread further apart, but the "bigness" of the work will still shine through!

Quantity and Symbol Match - Dab and Trace Units

Directions: Count the beads and use a dot-marker to dab the beads as you count by units. Alternatively, you can color them "gold." (You can use brown or orange. The original "Golden Beads" were orange!) Then, trace the number.

5
6
7
8
9

Quantity and Symbol Match - Dab and Trace Tens

Directions: Count the beads and use a cotton swab dipped in paint to swipe the length of each bead bar as you count by tens. (Alternatively, color the bead bars as you count by tens.) Then, trace the number.

10

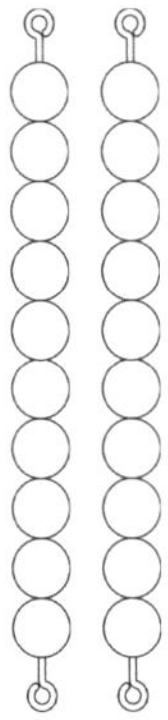

20

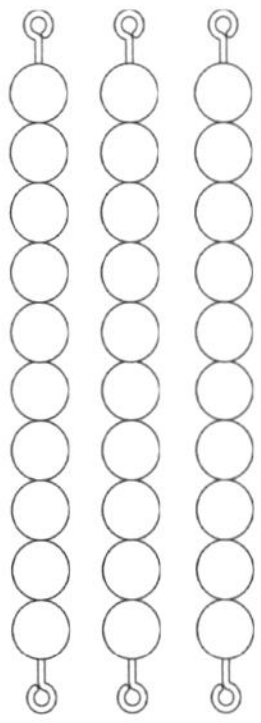

30

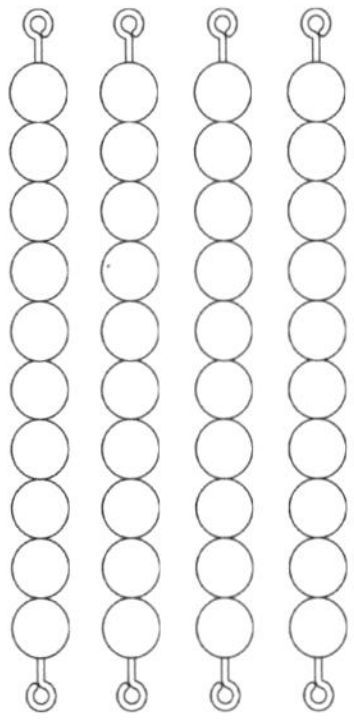

40

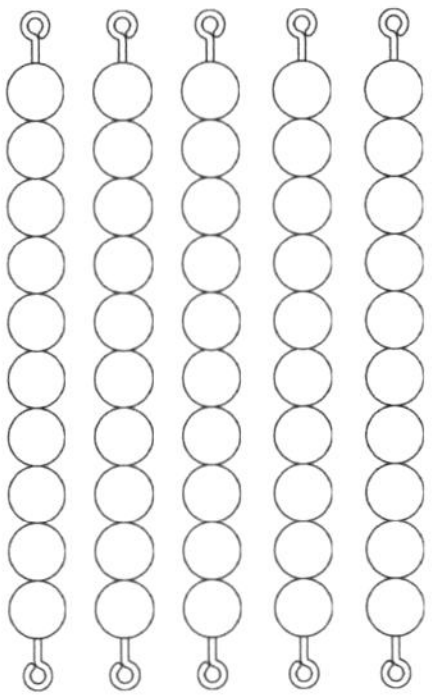

50

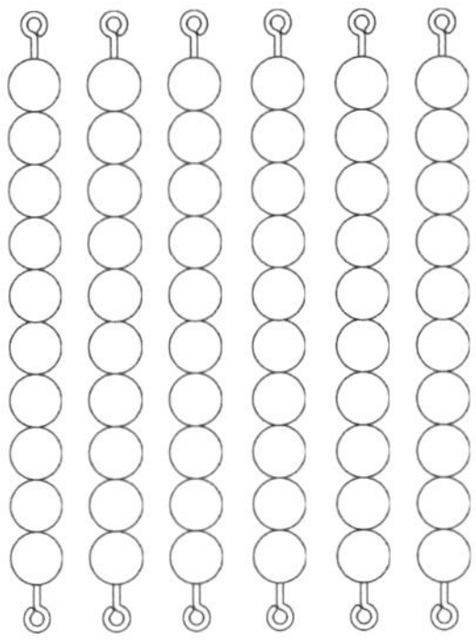

60

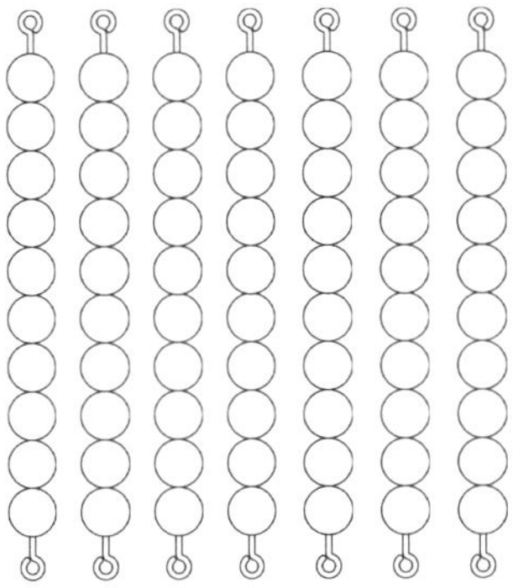

70

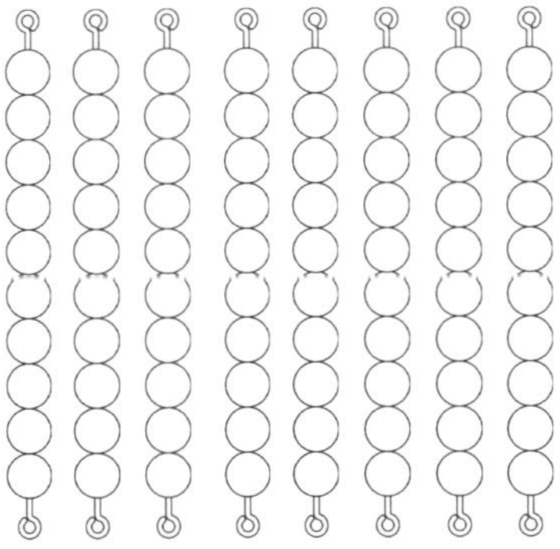

80

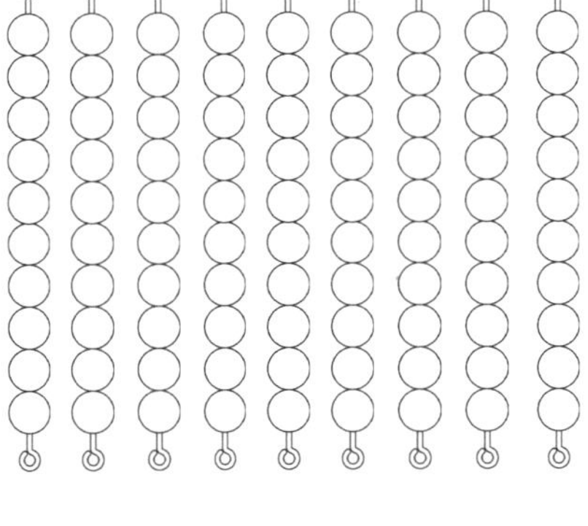

90

Quantity and Symbol Match - Dab and Trace Hundreds

Directions: Count the beads and use a dot-marker to dab the hundred squares as you count by hundreds. (Use one "dab" for each hundred square.) You can also color them "gold." (You can use brown or orange. The original "Golden Beads" were orange!) Then, trace the number.

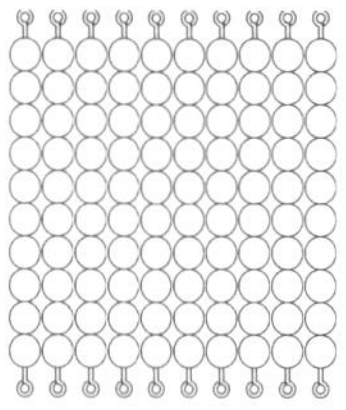

100

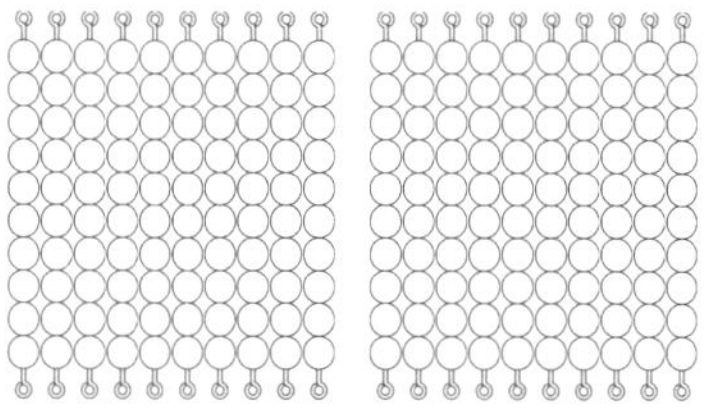

200

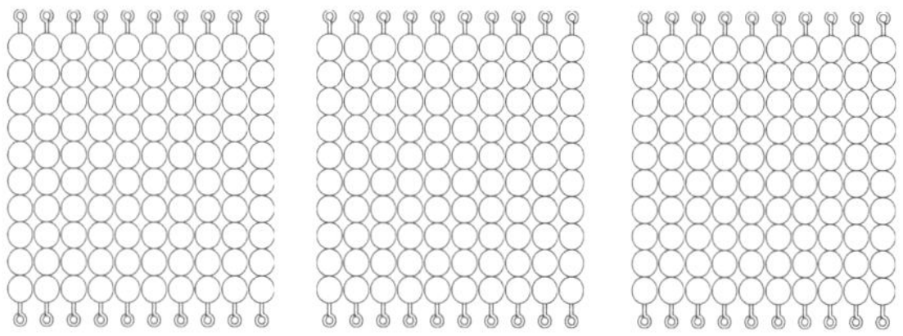

300

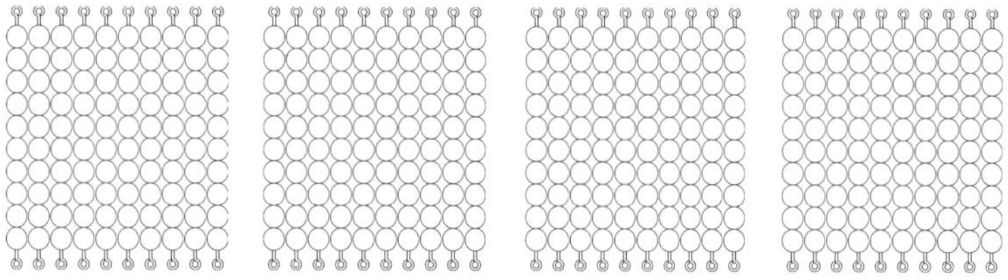

400

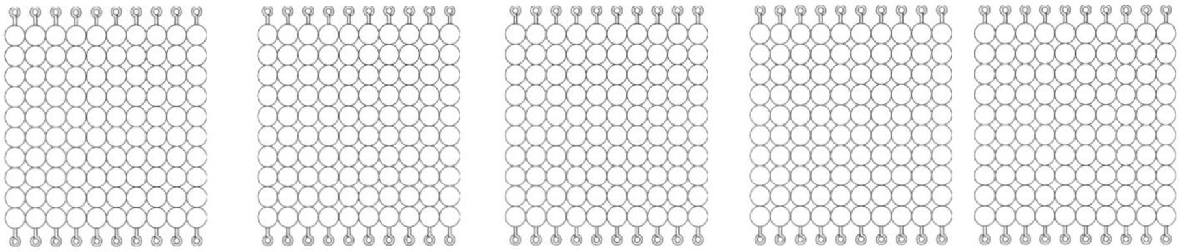

500

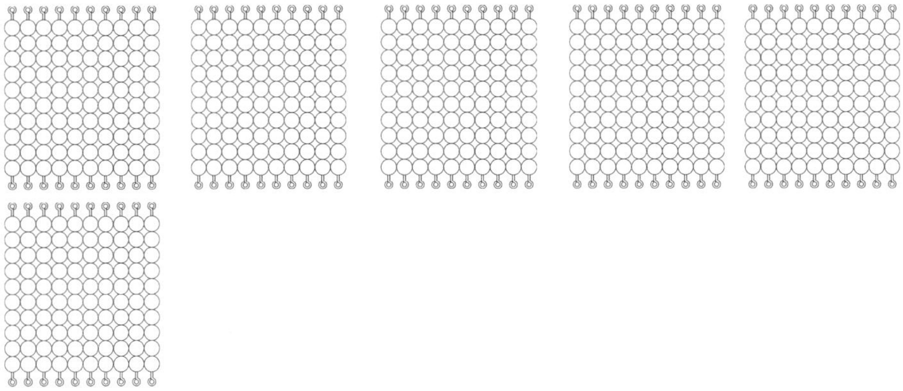

600

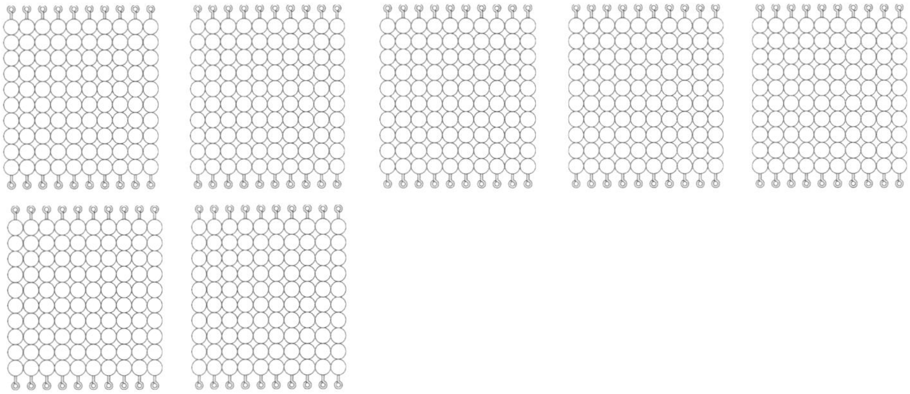

700

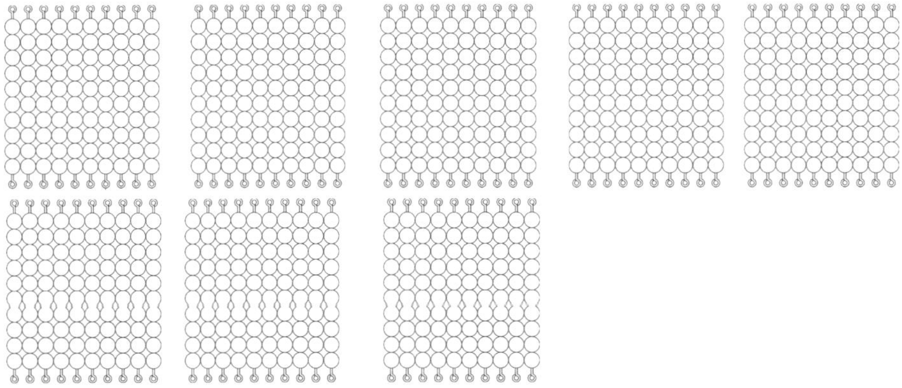

800

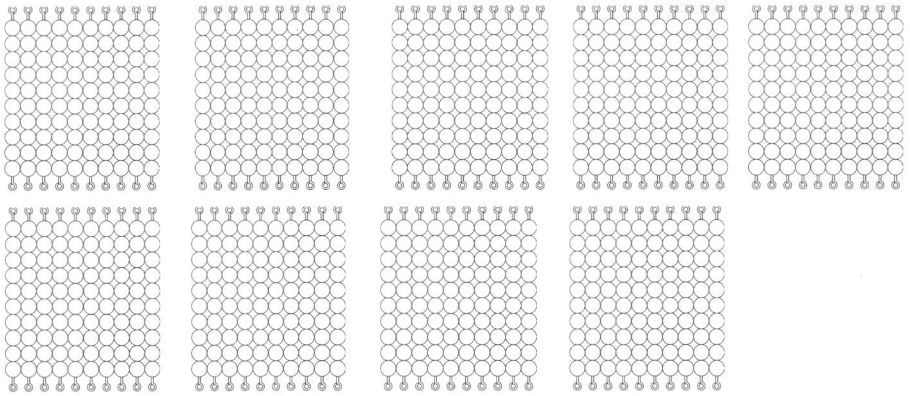

900

Quantity and Symbol Match - Dab and Trace Thousands

Directions: Count the beads and use a dot-marker to dab the cubes as you count by thousands. (Use one "dab" per cube.) You can also color them "gold." (You can use brown or orange. The original "Golden Beads" were orange!) Then, trace the number.

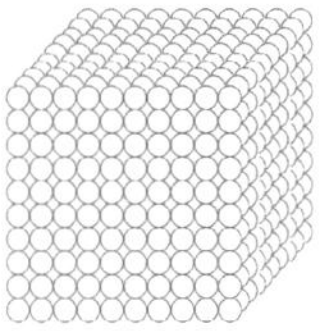

1000

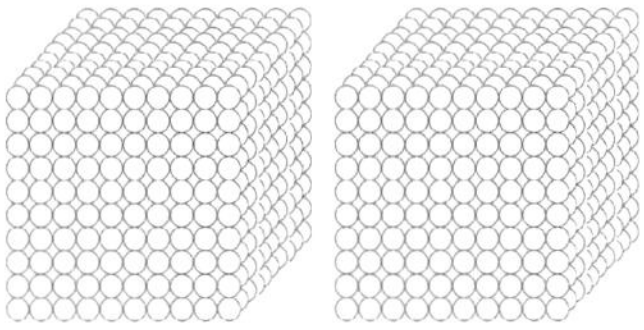

2000

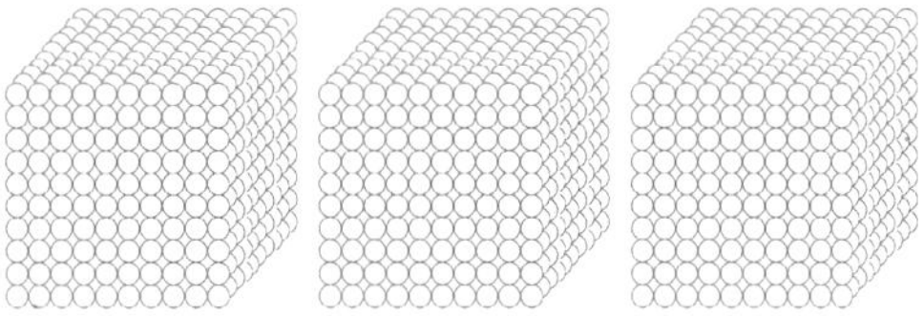

3000

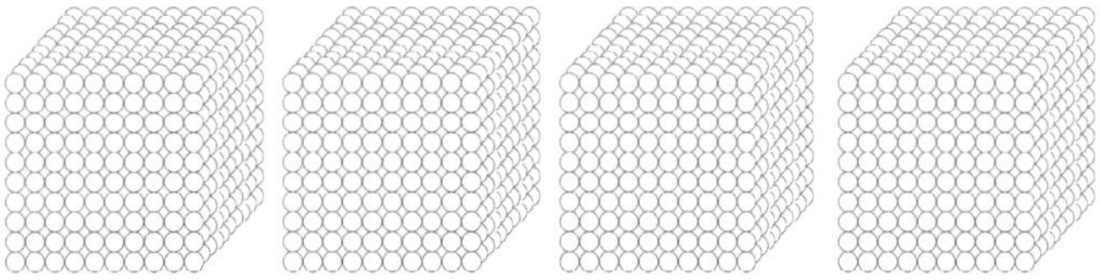

4000

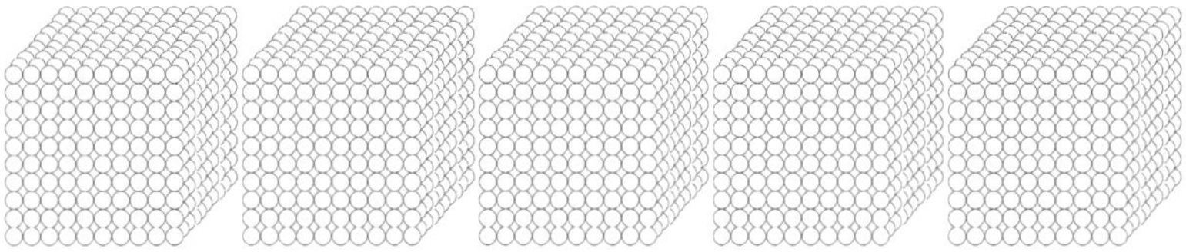

5000

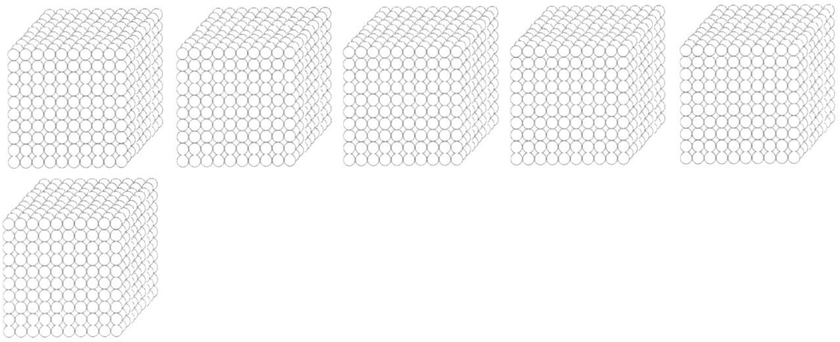

6000

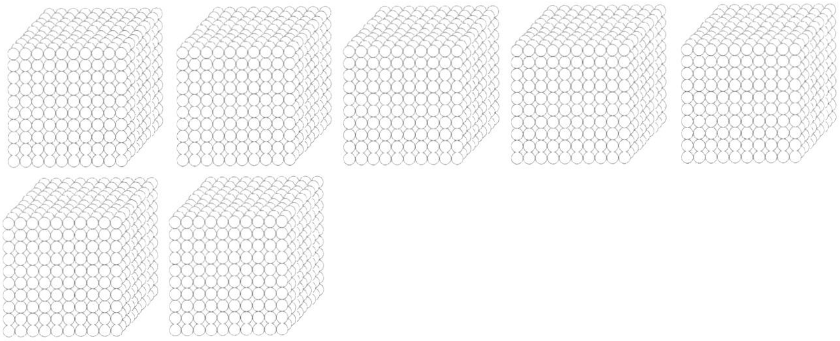

7000

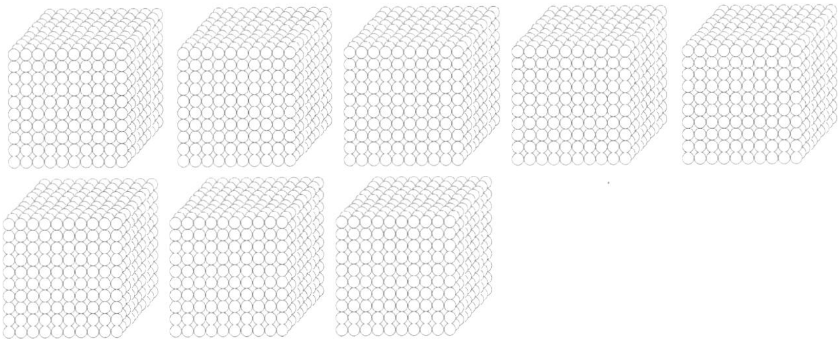

8000

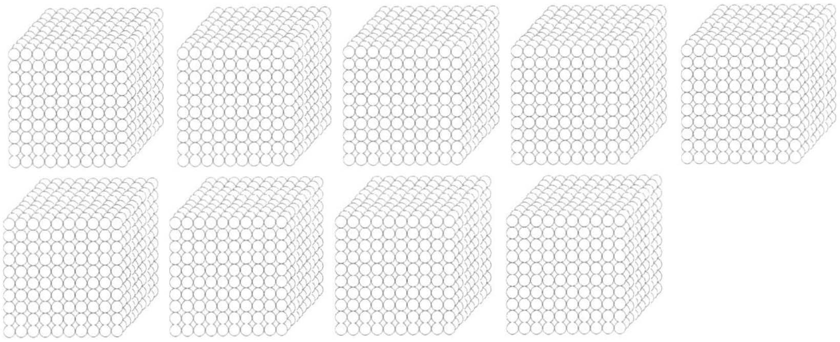

9000

Crises of Ten Lesson

Directions: You are going to show your child that there is not ever any more than nine of one category in the decimal system. Using the bead cards on the next page, use ten cut-out bead cards of each unit: thousands, hundreds, tens, and units. You will **not glue these down**. You will re-use them for the "Given the Symbol, Fetch the Beads" lesson.

On the "Crises of Ten" page, you will find the laid out numeral cards.

1. Point to the "1." Say, "This says one." Place one unit bead to the right of the "1." Say, "one."
2. Point to the "2." Say, "This says two." Move the unit bead down from next to the 1 and add one more unit bead. Count, "One, two."
3. Point to the "3." Say, "This says three." Move the unit beads down from next to the 2 one at a time and say, "One, two," and add one more unit bead and say, "Three."
4. Continue all the way down to the 9.
5. Point to the "10." Say, "This says ten." Move all of the beads up to next to the "10" in the next column, counting them aloud as you add. When you get to ten, *exchange the ten unit beads for one bead bar.* Say, "This is one ten."
6. Keep moving down the tens, counting aloud, and "make the exchange" after placing ten ten-bars next to 100. *Make sure to count the tens as "two tens, three tens, four tens, five tens, etc.
7. Keep moving down the hundreds, counting aloud, and "make the exchange" for the one thousand cube after placing ten hundred squares next to 1000. *Make sure to count the hundreds as one hundred, two hundreds, three hundreds, etc.
8. Count the thousands all the way to nine thousands. *Make sure to call them, "one thousand, two thousands, three thousands, etc."

Beads for "Crises of Ten Lesson" and "Given the Symbol, Fetch the Beads"

Crises of Ten Lesson Board

1000	100	10	1
2000	200	20	2
3000	300	30	3
4000	400	40	4
5000	500	50	5
6000	600	60	6
7000	700	70	7
8000	800	80	8
9000	900	90	9

Go Fetch: Decimal System

Directions: On the following pages, you will find two sets of problems: "Given the Symbol, Fetch the Beads," and "Given the Beads, Fetch the Numeral."

Using the cut-and-paste bead cards and decimal numeral cards, lay the full set of beads or cards out on the work surface. *This activity works well to incorporate movement and memory. After neatly laying out the beads or cards, invite your child to move to another area with the workbook and have him "fetch" the needed beads or cards by memory. Make sure the workbook stays at the distant location.

Demonstrate for your child how to read each box, starting with the units. Glue in the proper amount of beads or the proper numeral card in each box.

Show your child the examples below.

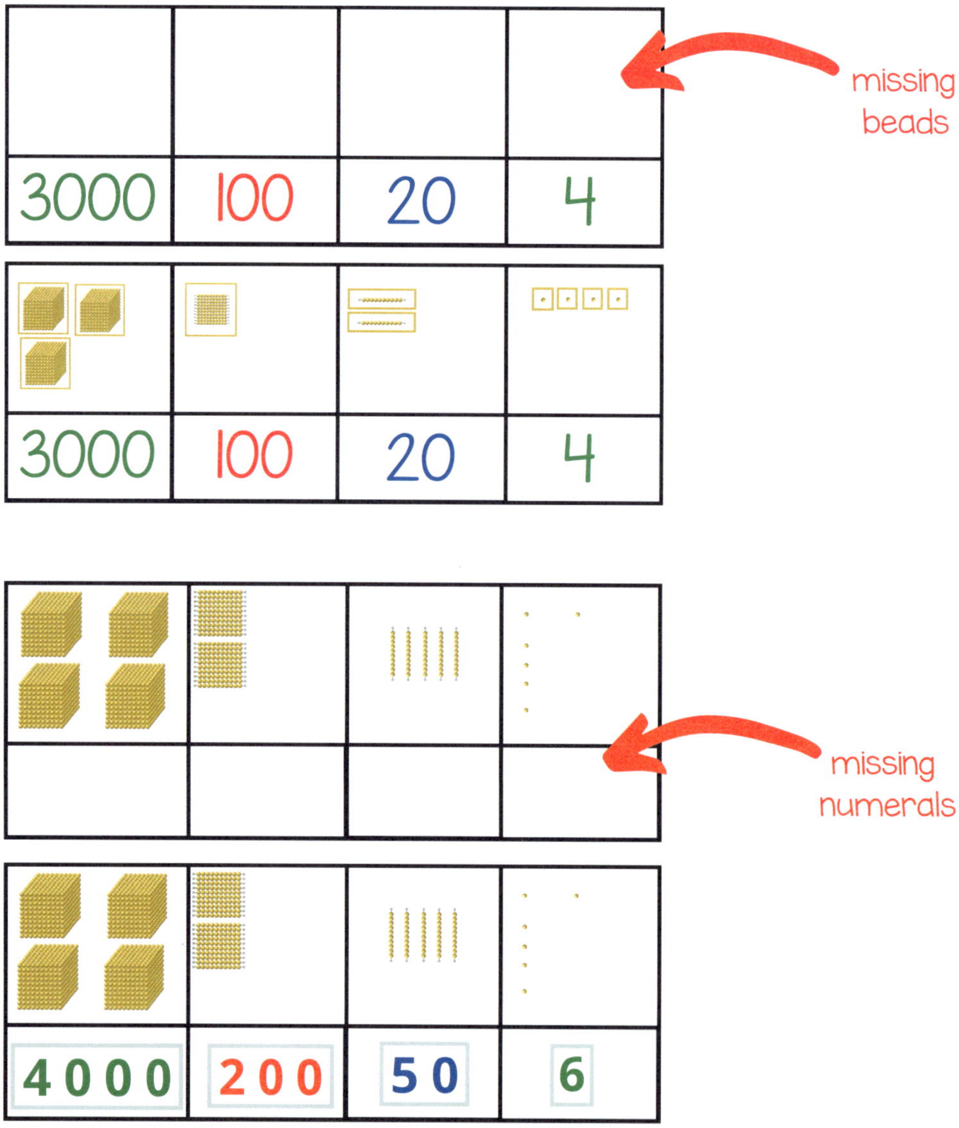

Numeral Cards for "Given the Beads, Fetch the Numeral"

1000	100	10	1
2000	200	20	2
3000	300	30	3
4000	400	40	4
5000	500	50	5
6000	600	60	6
7000	700	70	7
8000	800	80	8
9000	900	90	9

Given the Symbol, Fetch the Beads

4000	600	30	2

1000	500	0	6

Given the Symbol, Fetch the Beads

7000	300	40	3

2000	400	60	0

Given the Symbol, Fetch the Beads

0	500	20	4

3000	0	90	1

Given the Beads, Fetch the Symbol

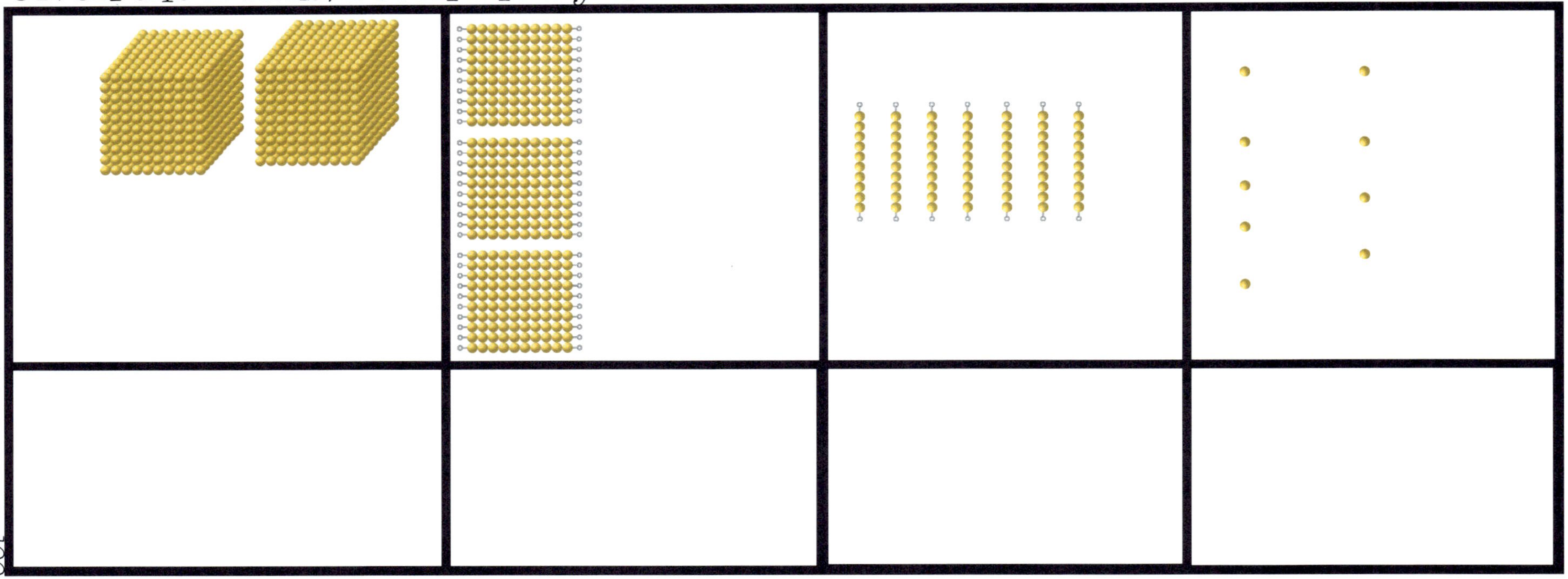

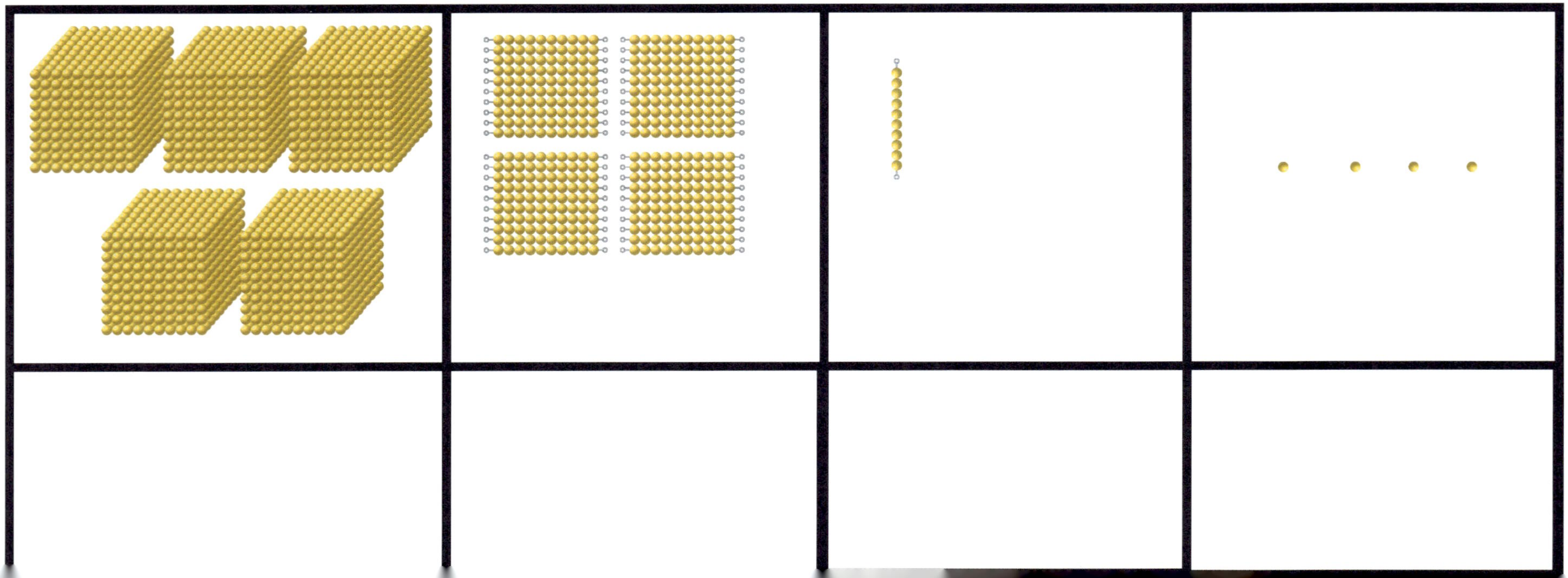

Given the Beads, Fetch the Symbol

Given the Beads, Fetch the Symbol

Formation of Numerals Intro

Directions: In this lesson, you will teach your student how to form numerals using the large numeral cards you have saved from the "Decimal System Cards - Full Layout" activity. Lay out the large numeral cards as shown before. After following the directions in the problem, write the numerals in the boxes or invite your student to write the numerals, using the matching colors of colored pencils. Then replace the cards to their places.

1. "Please bring me the numeral cards for five tens and eight units." Place the cards inside the box with the 8 on top of the 5. Then, lift them up, stack them vertically, and tap them into place so that the right ends align and the 8 covers the 0 in 50. Then, lay them back down and place them in the box. Read the number to your child, "fifty eight."
2. "Please bring me seven hundreds and three tens." Align the 3 on top of the 7. Then pick them up and tap the edge touching the zeros on the table to slide the cards down. Lay them down, and place them in the box. Say, "Three hundred seventy." Say, "There is no unit over this zero. The zero is a placeholder for the unit. It saves its place." Make sure your student records the 0.
3. "Please bring me four thousands and two hundreds." Align the 4 and 2, then tap them into place on the table and set them back down. Read the number, "Four thousand two hundred." Say, "There are two zeros here. This zero holds the place for the tens, and this zero holds the place for the units. The zeros are placeholders."
4. "Please bring me one thousand, four hundreds, and five tens." Align the 1, 4, and 5, then pick up the stack of cards and tap them into place. Set them in the box and read the number, "One thousand, four hundred, fifty." Say, "This zero is a placeholder for the units."
5. "Please bring me nine hundreds, eight tens, and two units." Align the 9, 8, and 2, then pick them up and tap the cards into place. Set them inside the box and read the number, "Nine hundred eighty two."
6. "Please bring me two thousands, six tens, and three units." Align the 2, 6, and 3, then pick the cards up and tap them into place. Set them inside the box and read the number, "Two thousand sixty three." As you read, skip over the zero, but pause there. Then point to the zero and say, "This zero is a placeholder for the hundreds."

1

2

3

4

5

6

Formation of Numerals

Directions:

1. Cut out all of the numeral cards on the following page.
2. Lay them out on your work space the same way they were arranged before you cut them.
3. Read the command to retrieve each set of number cards.
4. Invite your child to stack them by aligning them on the right hand side. (Since they are small and made of thin paper, it will be hard to "tap" them into place as you did with the larger, laminated numeral cards.)
5. Read the number to your child after it is formed as a single number. (For example, 8647 will be "eight thousand six hundred forty seven.")
6. Glue number down.

Eight thousands, six hundreds, four tens, and seven units.

One thousand, two hundreds, three tens, and five units.

Two thousands, eight hundreds, and seven tens.

Nine thousands, five hundreds, and four units.

Five thousands, seven hundreds, nine tens, and two units.

Three thousands, four hundreds, eight tens, and three units.

Seven thousands, one hundred, six tens, and one unit.

Four thousands, nine hundreds, two tens, and nine units.

Three hundreds, five tens, and eight units.

Six thousands, one ten, and six units.

Numeral Cards for "Formation of Numerals"

1000	100	10	1
2000	200	20	2
3000	300	30	3
4000	400	40	4
5000	500	50	5
6000	600	60	6
7000	700	70	7
8000	800	80	8
9000	900	90	9

Golden Beads Operations Overview

All of the work learning the decimal system will now be put to use as your child learns to add, subtract, multiply, and divide four digit numbers using the Golden Bead Material.

Materials Preparation:

1. **Golden Beads:** In these lessons, you will use one full set of Golden Beads (45 units, 45 tens, 45 hundreds, 9 thousands.) Your options for the Golden Beads are the same as for previous exercises. If you do not own a set of physical Montessori Golden Beads, there are three other options in this workbook (if you have worked in order, you should have your Golden Beads saved to use here):
    a. Option A - use the printables to make a full set of three-dimensional Golden Beads.
    b. Option B - Use the printables to make a set of large Golden Bead Cards.
    c. Option C** - Use the small set of Golden Bead Cards. (These work for the Operations Mats included in the workbook!)
2. **Numeral Cards:** You need one set of large numeral cards (you should already have these), and two or three sets of small numeral cards, which are found on pages 93, 95, and 97 of the Workbook. Cut out and laminate these!
3. **Operations Mats**:** Each operation has its own "mat" in this workbook. You may want to remove these pages and laminate them or place them in plastic sleeve protectors. Make sure to tape them together to make a single mat. (Note: The Division Mat consists of four separate mats. Look at the instructions page for an image of what it should look like.)
   **The only Golden Beads that fit on these operations mats are the small Golden Beads Cards (Option C). If you are using real Golden Beads or the printable 3-Dimensional Option, you can make your own mats using Felt and Hot Glue or even Construction Paper taped together! You can also choose to do the operations with no colored mats. You can do them on a table top or plain work rug.

Pacing the Lessons

There are 6-12 problems for each lesson type. I present one full problem as an example for each type of problem. I suggest showing your child the example problem fully as you look at how it's presented in the introduction. Then, encourage your child to do one herself. If your child shows interest, do more! You do not have to do all the problems in each section, *especially not in one sitting*. Time how long it takes your child to do one problem, then set a reasonable goal for the day and each day moving forward as you work through the book. Completing 1-2 problems in one sitting is an excellent goal for a 4 or 5 year old!

Notes and Definitions

Static means "not changing."
Dynamic means "changing." In Dynamic operations, we are going to have to ex-"change" one unit of bead for another. First, we will remind our children that there can never be more than nine in any one category in the decimal system. We will help our children "make the exchange" with ten of one category for one of the next higher category. (Example: 10 unit beads exchanges for one ten-bar.)
Addition - Add two (or more) **addends** together to obtain the **sum**. The color for addition is red.
Subtraction - Subtract the **subtrahend** from the **minuend** to obtain the **remainder**. The color for subtraction is green.
Multiplication - Multiply the **multiplicand** by the **multiplier** to obtain the **product**. The color for multiplication is yellow.
Division - Divide the **dividend** by the **divisor** to obtain the **quotient**. The color for division is blue.

Golden Beads Static Addition

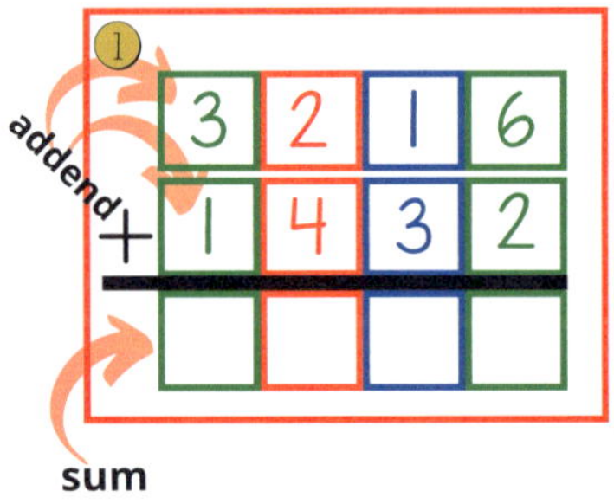

1

Look at the problem. Fetch one set of **small numeral cards** at a time. Then lay them out on top of each other, to the left of your "work space." Read the addends as you place them, "Three thousand, two hundred, sixteen."

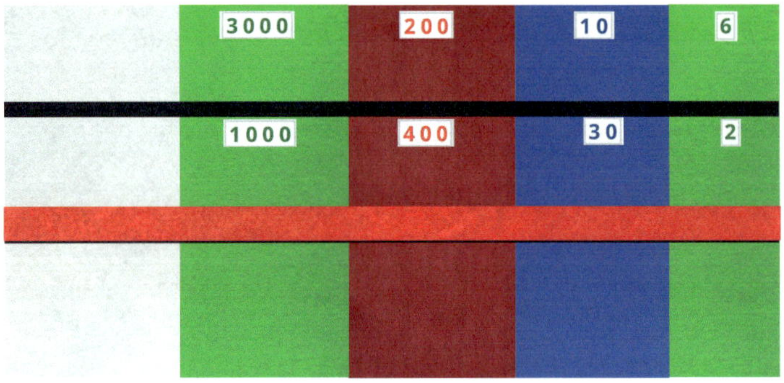

2

Separate out the **small numeral cards** to their decimal places.

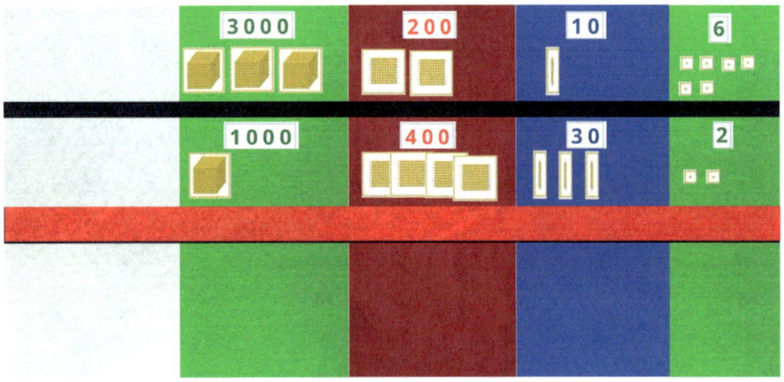

3

Fetch the proper number of beads for each place value. First fetch the thousands, then the hundreds, then the tens, then the units for the first addend. Repeat for the second addend.

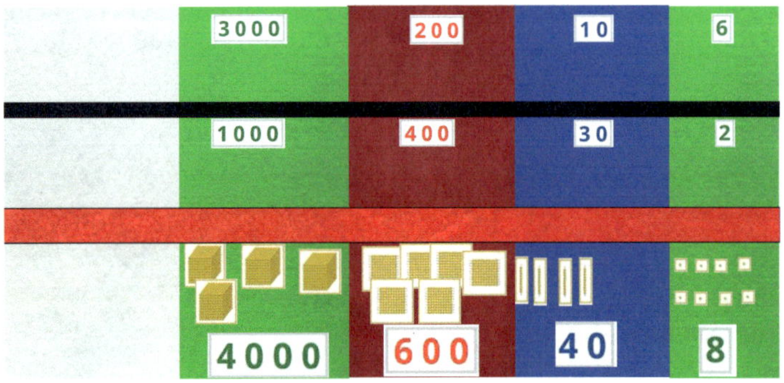

4

Say, "Now it is time to add all of our golden beads together. We always start with the units." Combine the units by bringing all of them down to the bottom row. Repeat for the tens, hundreds, and thousands. Then count the units, and fetch the matching **large numeral card**. Repeat for other values.

5

Combine the small numeral cards into one number, as shown before, for both addends. Combine the large numeral cards as shown before for the sum. Say the problem out loud: "Three thousand, two hundred, sixteen plus one thousand four hundred thirty two equals four thousand six hundred forty eight."
(Point to the addends and say, "This is an addend." Point to the sum and say, "This is the sum.")

6

Using the proper colors of colored pencils, invite your child to copy in the answer on the problem grid.

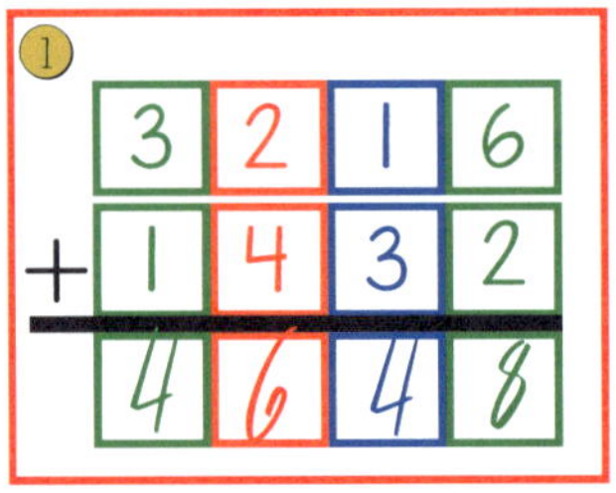

Golden Beads Static Addition Problems

1

	3	2	1	6
+	1	4	3	2

2

	2	5	3	7
+		4	2	1

3

	5	3	4	1
+	2	1	4	8

4

	5	1	2	4
+	3	0	6	0

5

		1	5	7
+		8	2	0

6

	1	7	1	3
+	6	2	1	4

Golden Beads Static Addition Problems

7

	6	8	0	1
+	2	1	4	5

8

	5	2	1	0
+		1	0	6

9

	3	3	8	2
+	1	6	1	7

10

	8	1	9	0
+		5	0	3

11

	6	7	4	1
+	2	1	5	6

12

	2	2	3	5
+	2	1	0	3

Golden Beads Addition Mat

Remove the mat from the workbook to use alongside the addition problems. Laminate or place in page protectors for durability.

Tape the middle of the mat together.

*Note: If you are using real Golden Beads or the larger Golden Beads Printables, you may want to replicate this mat in a larger size with felt or colored paper.

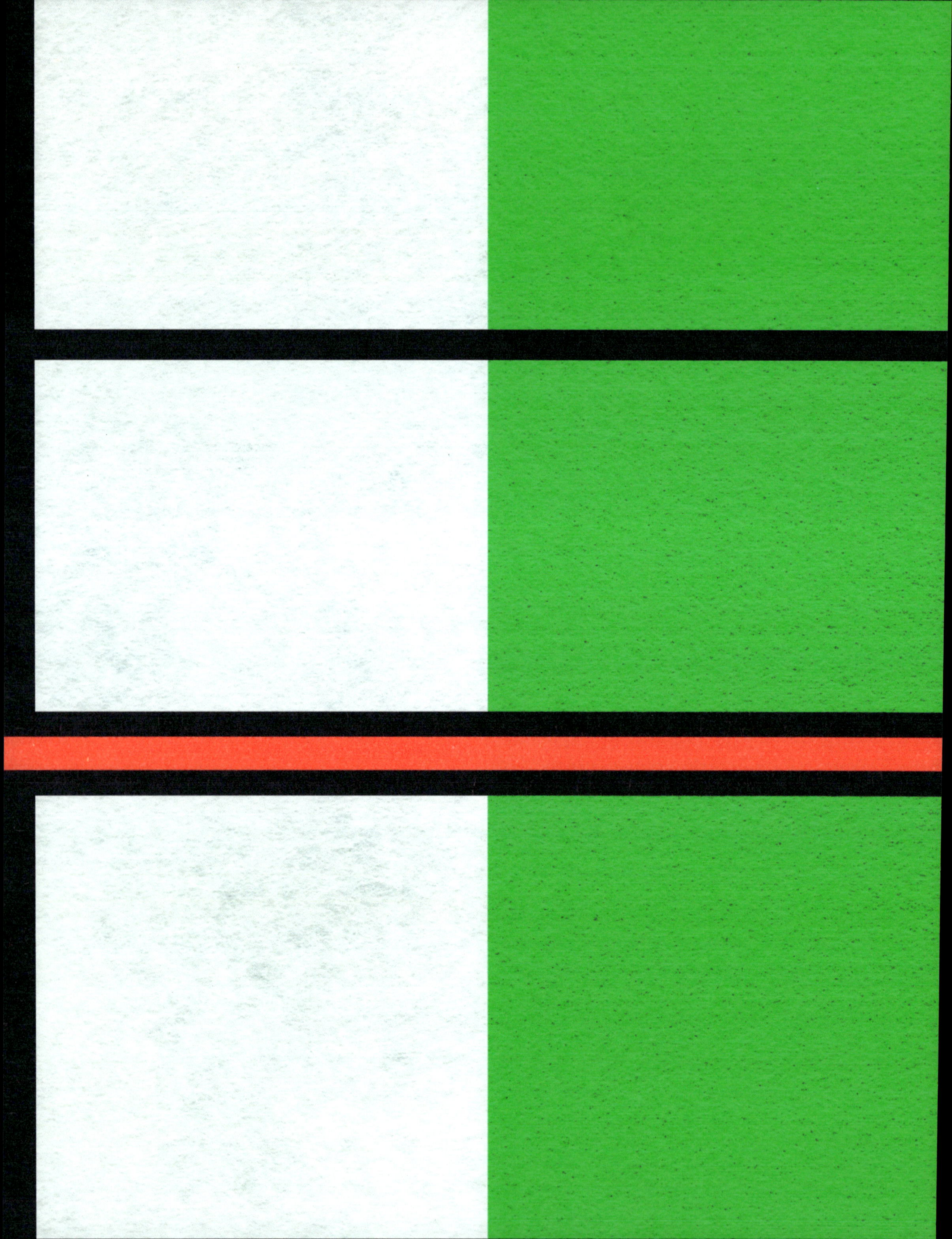

Matching Quantities

Match the quantities on the left to the quantities on the right by drawing a line from a box on the left to a box on the right..
Optional: Invite your child to write in the numeral representing the quantity in each of the boxes in the right column.

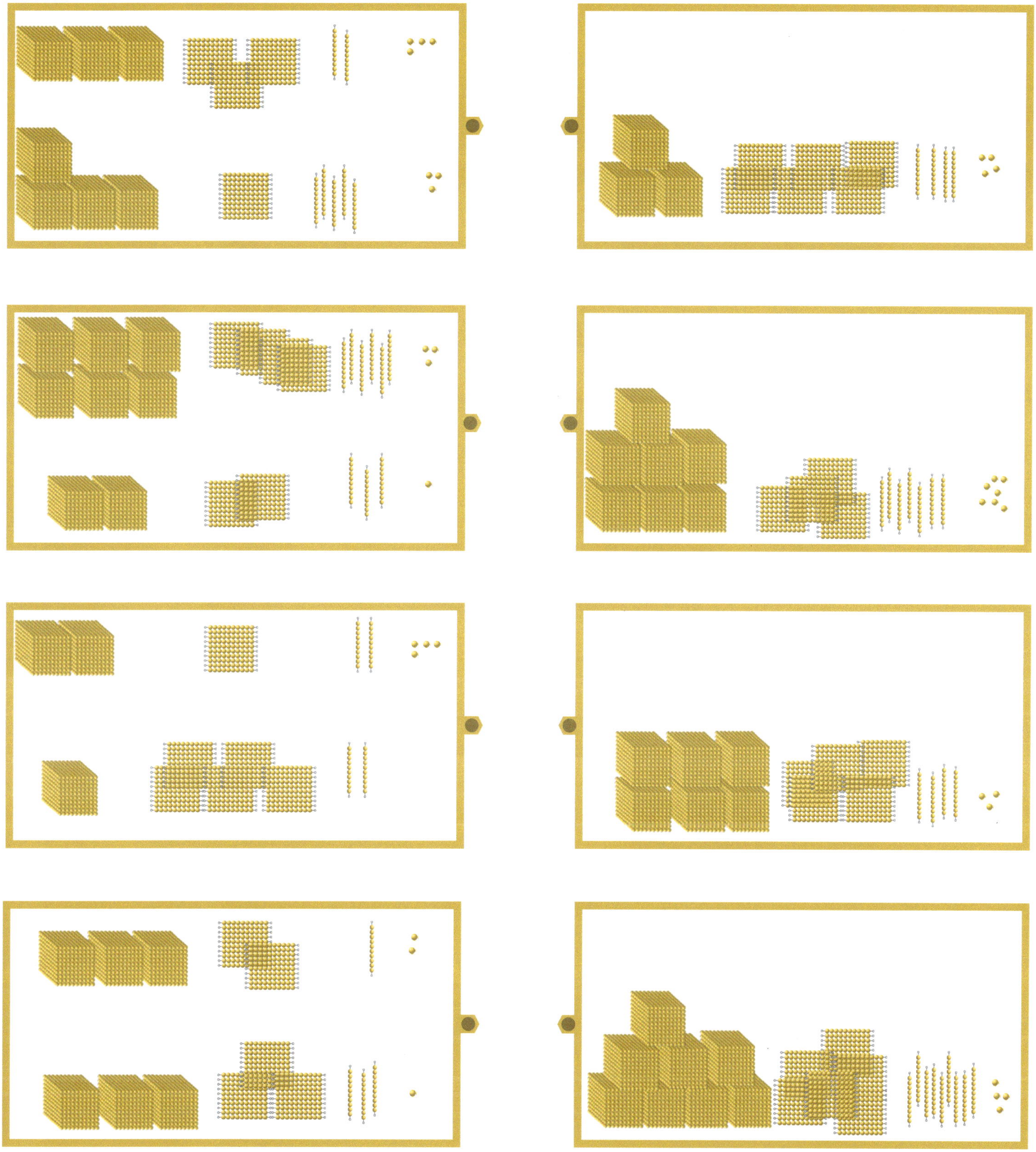

Exchanging for Ten

We are going to review exchanging. Explain to your child, "We can never have more than nine in any one category. When we do, we need to make a group of ten and exchange." Today, we will practice finding groups of ten and exchanging for the next category in the decimal system.

1. Circle groups of ten units.
2. "Exchange" each circled group of ten for one higher unit than the unit you circled. Cross out the circled group after exchanging.
3. Paste in the new version of the beads into the bottom box. If you have "leftovers," make sure to represent those beads by pasting new ones in the lower box, as well.
4. Write the number in the box using the correct colors of pencils (match the decimal system cards.)

Answers:

1. For the first box, you will paste 10 ten-bars and 5 unit beads.
2. Paste two hundred squares and one ten-bar.
3. Paste one thousand cube and two hundred squares.

Note: You will not use all available beads that you cut out.

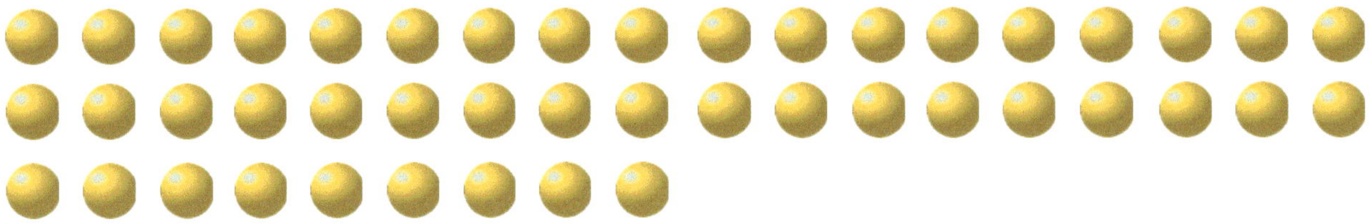

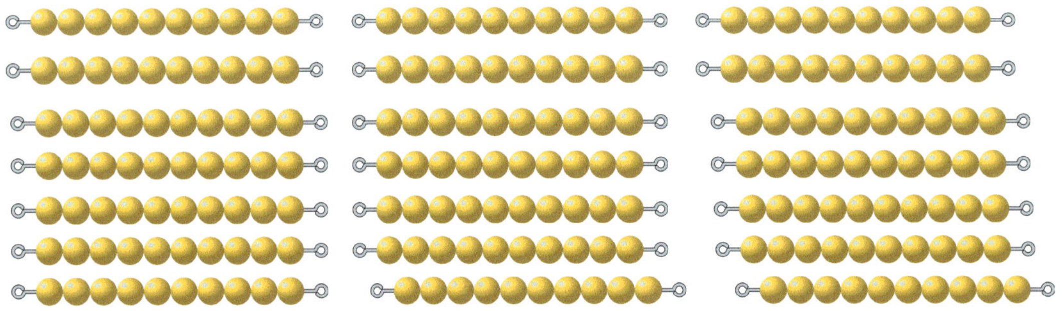

Exchanging for Ten

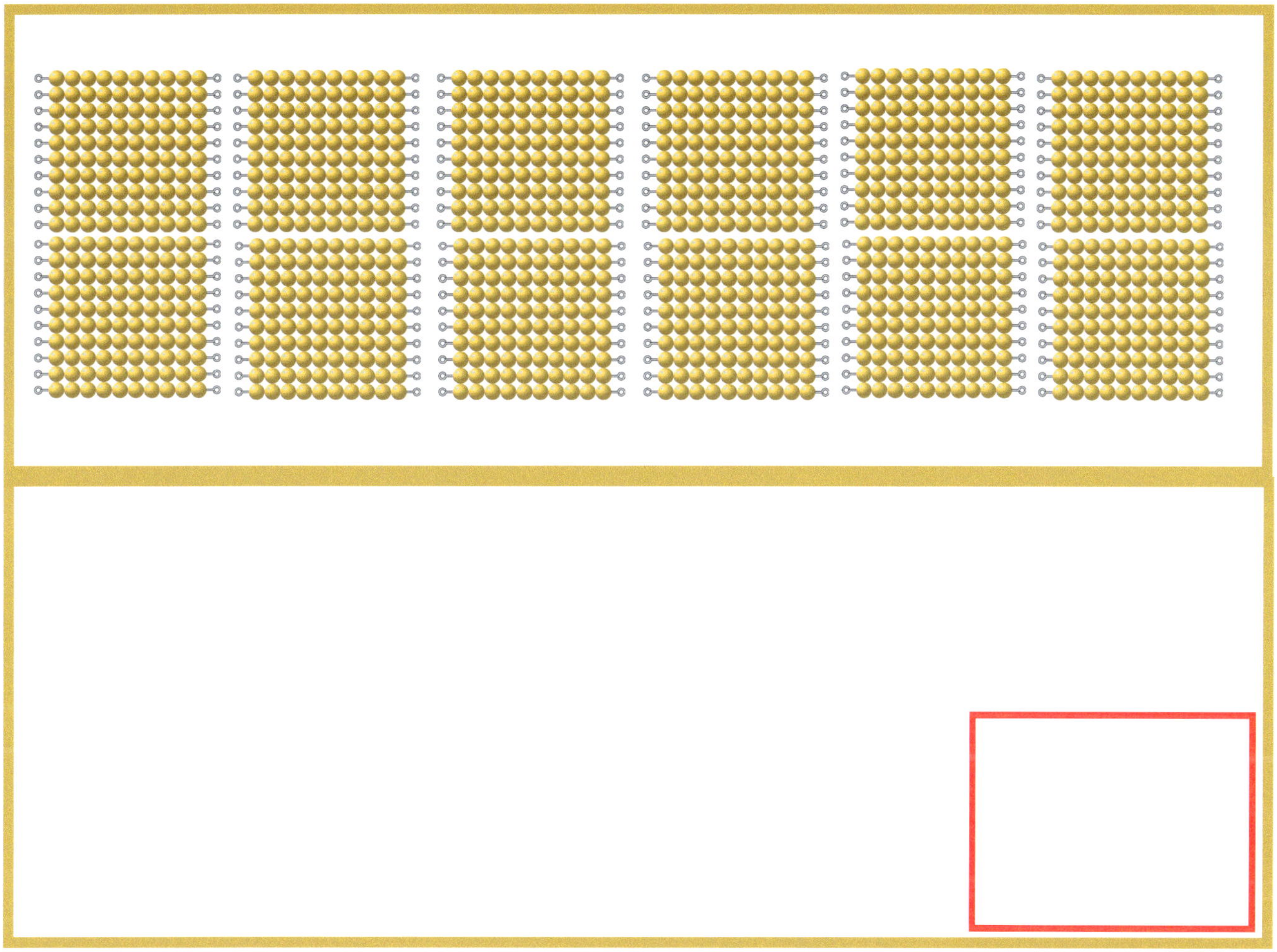

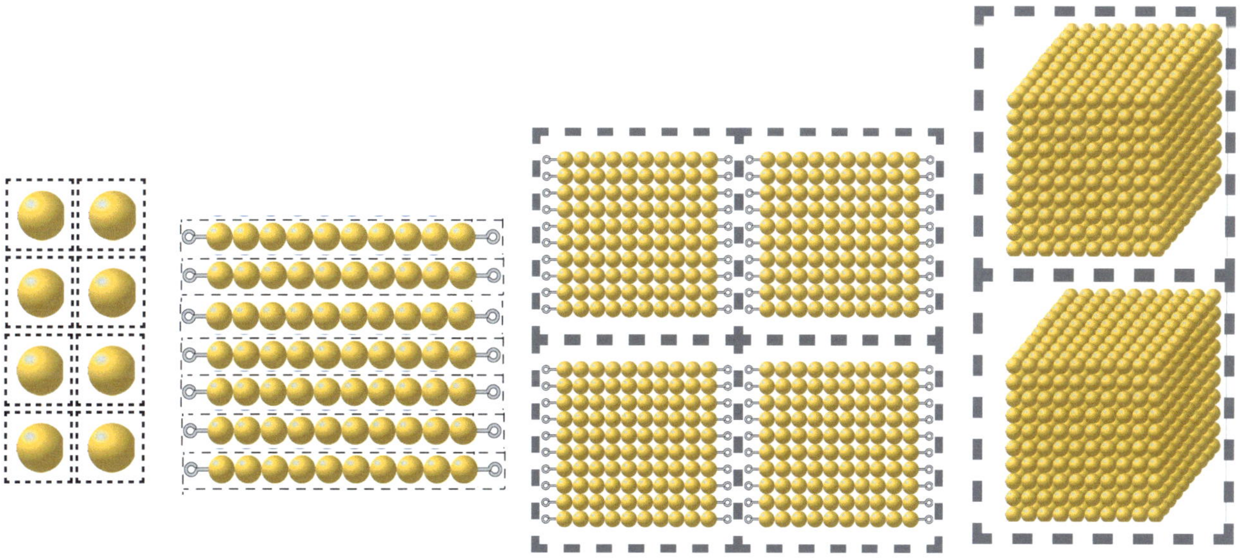

Golden Beads Dynamic Addition

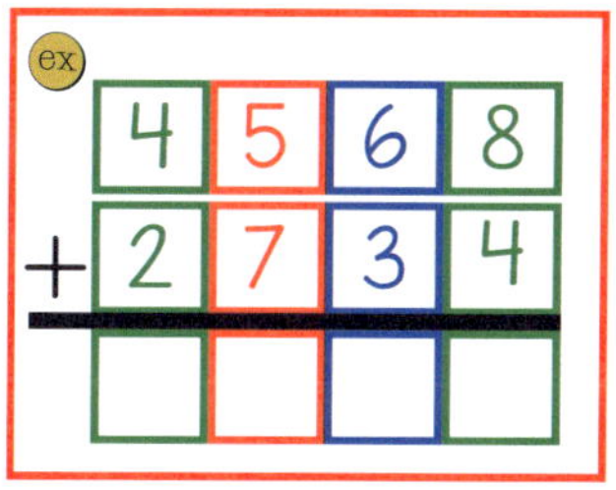

1

Look at the problem. Fetch one set of **small numeral cards** at a time. Then lay them out on top of each other, to the left of your "work space." Read the addends as you place them, "Three thousand, two hundred, sixteen."

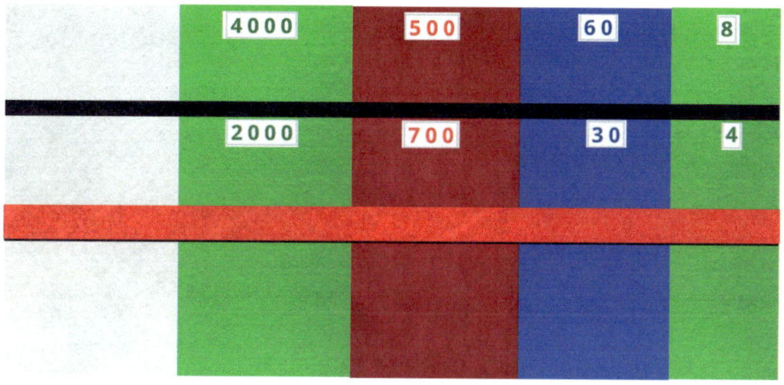

2

Separate out the **small numeral cards** to their decimal places.

3

Fetch the proper number of beads for each place value. First fetch the thousands, then the hundreds, then the tens, then the units for the first addend. Repeat for the second addend.

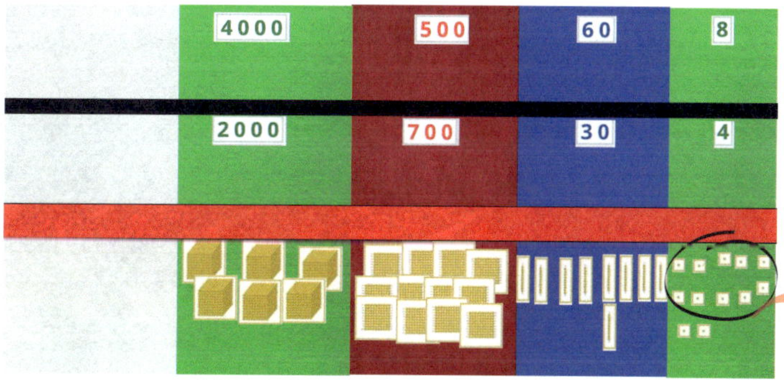

4

Say, "Now it is time to add all of our golden beads together. We always start with the units." Combine the units by bringing all of them down to the bottom row. Repeat for the tens, hundreds, and thousands. Then count the units. Say, "We can't have more than 9. We have 12. We need to exchange ten unit beads for a ten-bar." Count out ten units, and **exchange** for a ten-bar from your Golden Beads Bank.

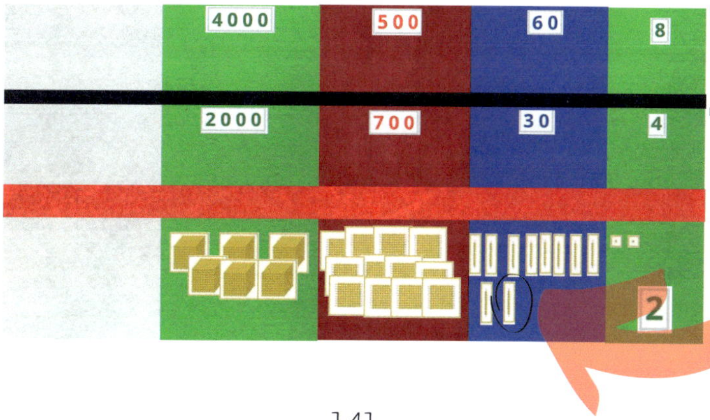

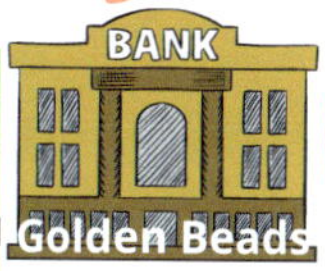

5

Add the ten-bar to the other tens. Re-count. leftover units and place the matching large numeral card below the unit beads..

6

Now count the ten-bars. If there are ten or more, make the exchange for a hundred square from the Golden Beads Bank. Count the remaining ten-bars.

7

Count the remaining ten-bars. **(If there is zero in any category, don't place any number.)**

Note: the new hundred square was added to the section of hundreds.

8

Count the hundred squares. If you get to ten, group those and exchange them with one thousand cube from the Golden Beads Bank.

9

Place the thousand cube in the thousands place on the mat or work space. Count the remaining hundred squares. Place the large numeral card for the hundreds on the mat.

Count the thousand cubes, and place the matching numeral card underneath the cubes.

10

Restack the small numeral cards. Stack the large numeral cards. Read the full problem out loud.

11

Using the proper colors of colored pencils, invite your child to copy the large numeral card value into the boxes on the problem grid.

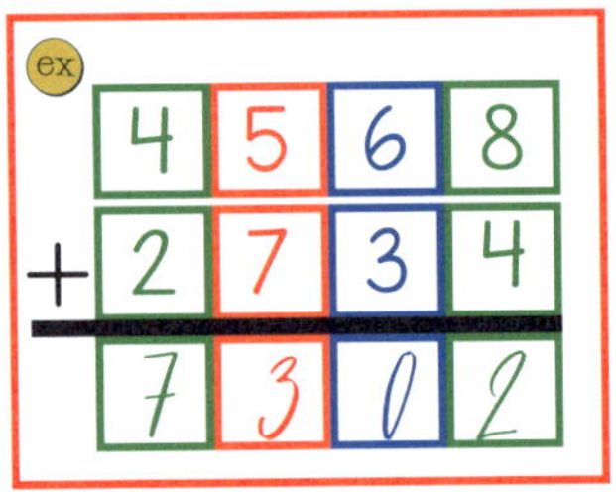

Golden Beads Dynamic Addition Problems

1

	3	1	7	8
+	1	8	5	3

2

	3	3	3	9
+		9	1	1

3

	5	7	6	4
+	2	5	5	9

4

	7	4	4	1
+	1	9	9	0

5

	2	1	3	9
+	5	9	3	1

6

	6	6	0	5
+		1	1	7

Golden Beads Dynamic Addition Problems

7

	4	1	1	6
+	5	3	1	1

8

	7	8	7	5
+		8	8	6

9

	1	7	6	0
+	1	3	7	6

10

	4	0	8	8
+	1	9	6	3

11

	2	9	0	3
+	2	8	4	4

12

	1	2	6	2
+	5	9	9	9

Golden Beads Subtraction Mat

Remove the mat from the workbook to use alongside the subtraction problems. Laminate or place in page protectors for durability.

Tape the middle of the mat together.

*Note: If you are using real Golden Beads or the larger Golden Beads Printables, you may want to replicate this mat in a larger size with felt or colored paper.

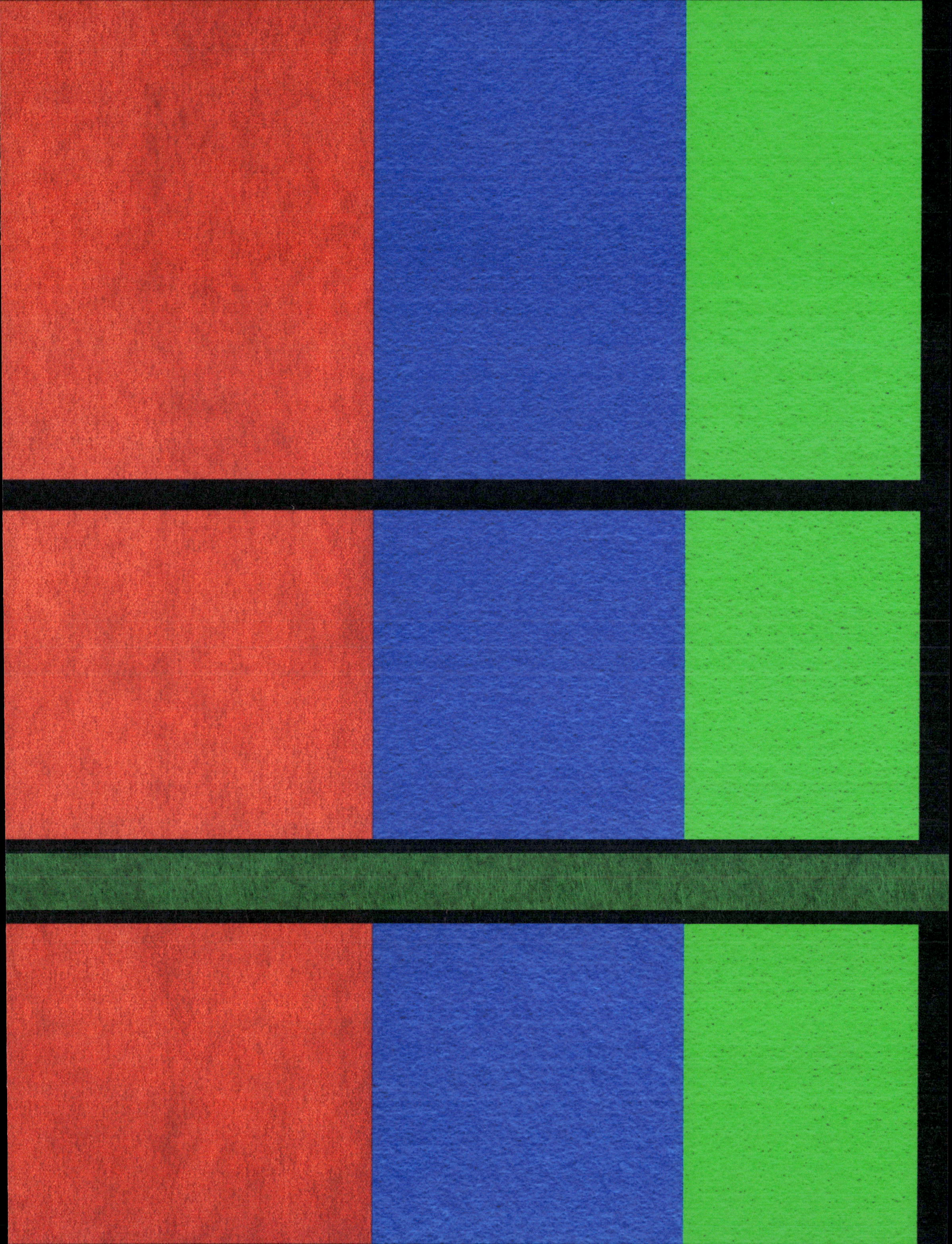

Golden Beads Static Subtraction

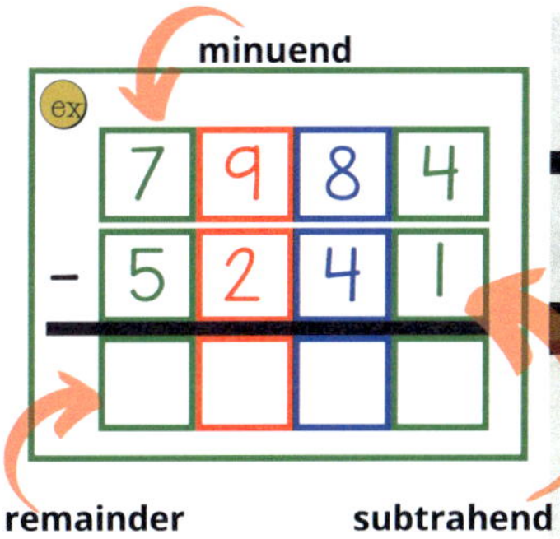

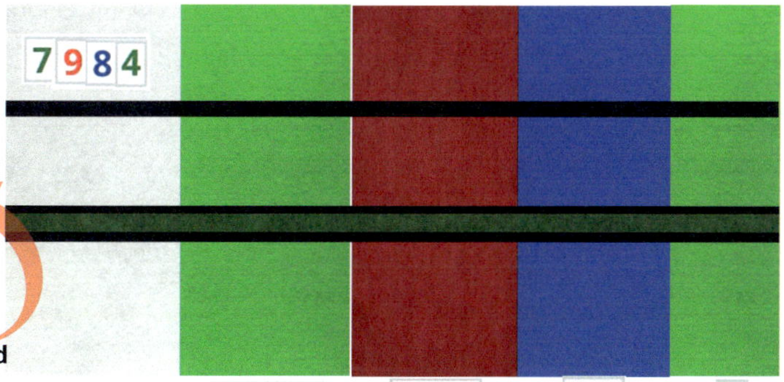

1

Invite your child to retrieve the **large numeral cards** for the minuend, and say the names of the values as they are placed on the left side of the work area. Say, "This is the minuend."

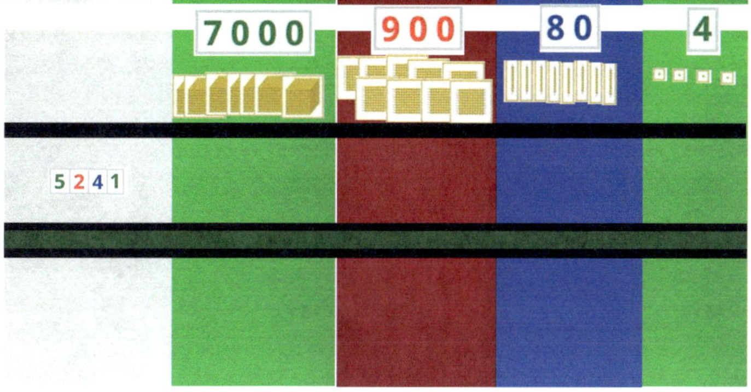

2

2a - Separate out the **large numeral cards** for the minuend, and say the name of the values as you place them.
2b - Next, invite your child to gather the Golden Beads for each category and lay them underneath the numbers.
2c - After laying the Golden Beads out, invite your child to retrieve the small numeral cards for the subtrahend. Read them as you or your child sets them down. Say, "This is the subtrahend."

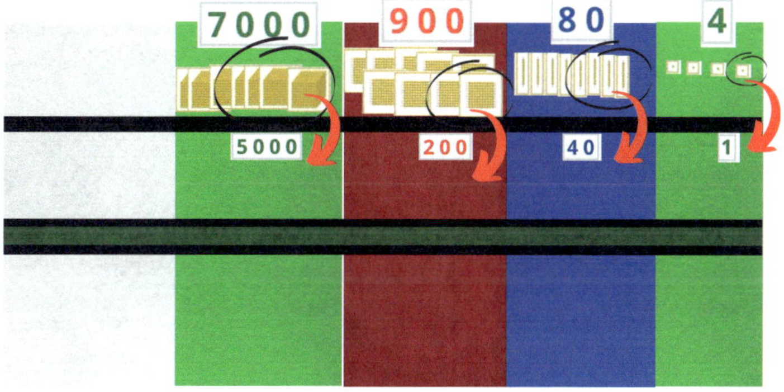

3

3a - Separate out the small subtrahend numeral cards, saying the names of the values as you set them down.
3b - Say, "We are going to **subtract** the **subtrahend** from the **minuend**." Show your child how to read the numeral card then move down the corresponding beads into the space below the small numeral cards.

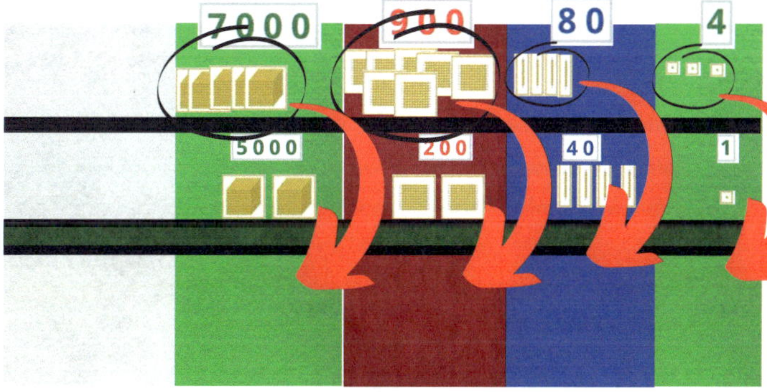

4

4a - The beads on the top row now need to be moved to the bottom row to show what is left over after subtracting the subtrahend from the minuend. Invite your child to move the beads down, starting with the units.
4b - Place a **small numeral card** for each category after moving the beads.

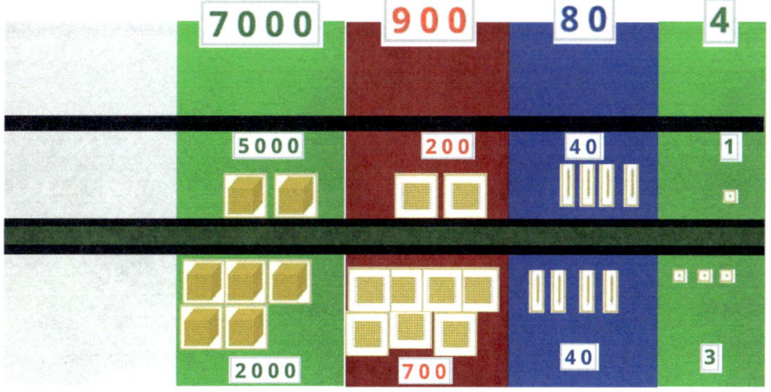

5

Lastly, stack all of the numeral cards again and place them on the left side (*not pictured.) Read the problem aloud. Say, "We started with 7984, the minuend. We subtracted 5241, the subtrahend. We were left with 2743 remaining, which is called the **remainder**."

6

Write in the remainder on the problem grid. Then replace all of the beads and the numeral cards.

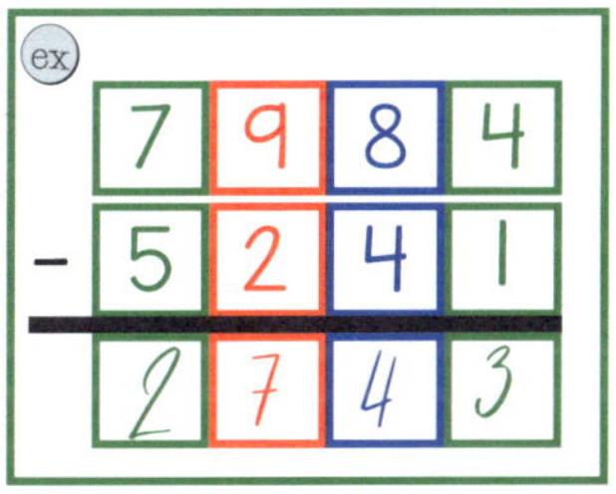

Golden Beads Static Subtraction Problems

1

	2	2	8	3
–	1	1	6	1

2

	8	5	1	8
–	6	2	1	3

3

	6	8	5	2
–	2	6	3	1

4

	1	5	9	3
–		4	2	2

5

	9	6	7	5
–	5	2	3	5

6

	2	2	8	3
–	1	1	8	2

Golden Beads Dynamic Subtraction

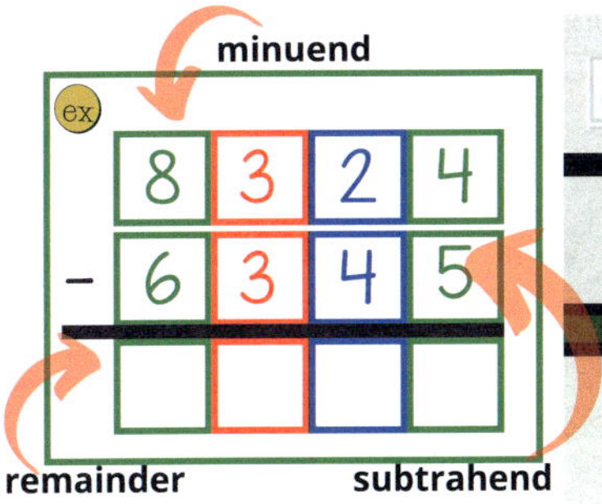

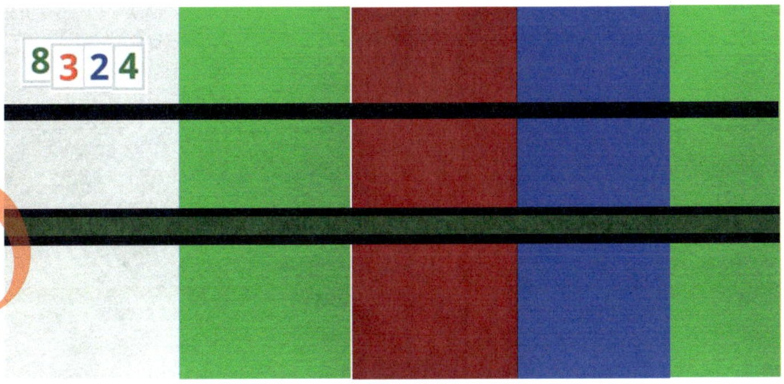

1

Invite your child to retrieve the **large numeral cards** for the minuend, and say the names of the values as they are placed on the left side of the work area. Say, "This is the minuend."

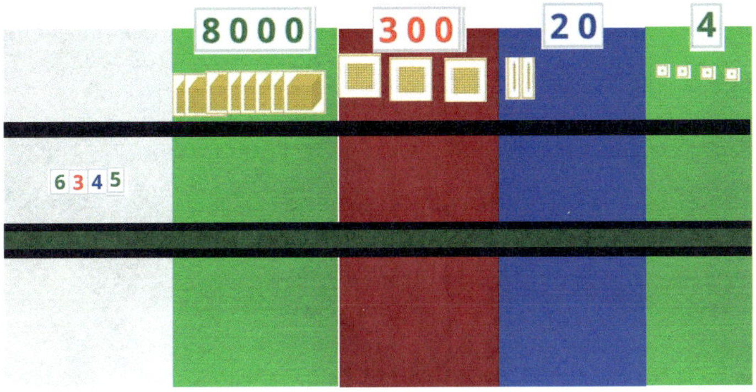

2

2a - Separate out the **large numeral cards** for the minuend, and say the name of the values as you place them.
2b - Next, invite your child to gather the Golden Beads for each category and lay them underneath the large numbers.
2c - After laying the Golden Beads out, invite your child to retrieve the small numeral cards for the subtrahend. Read them as you or your child sets them down. Say, "This is the subtrahend."

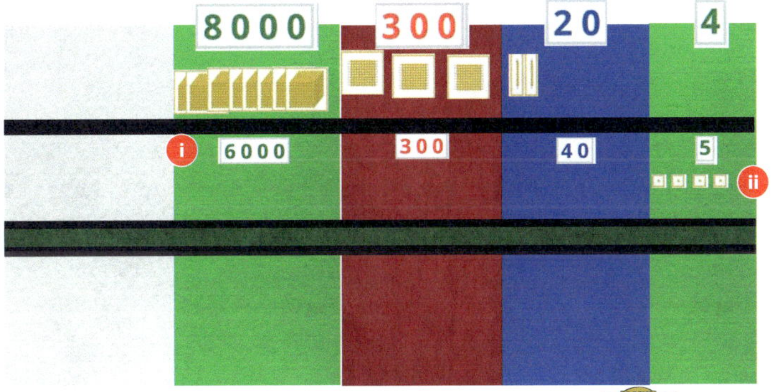

3

(i) Separate out the small subtrahend numeral cards, saying the names of the values as you set them down.
(ii) Say, "We are going to **subtract** the **subtrahend** from the **minuend**." Starting with the units, show your child how to read the numeral card then move down the corresponding beads into the space below the small numeral cards. As you place them, count the units.
(iii) If there are not enough units, say, "There are only four units. I need five units. I am going to take a ten from here (show your child) and exchange the ten-bar to ten units from the Golden Bead Bank!" Invite your child to make the exchange by placing the ten-bar in the bank and taking out ten unit beads. Count out the unit beads in the top units section to verify there are ten.
(iv) Then, re-count the beads you have already placed in the middle section, and take down the remaining needed units.

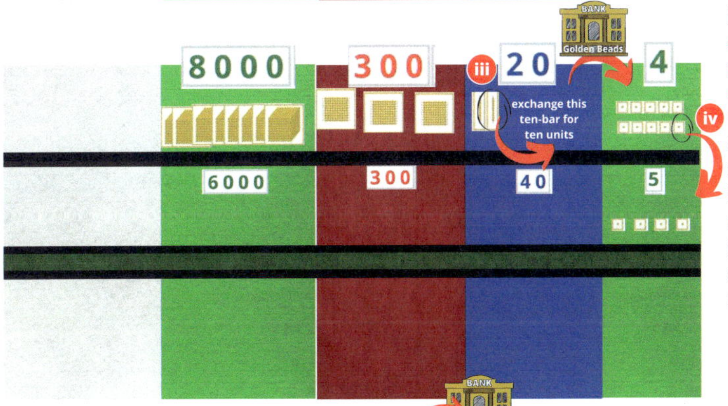

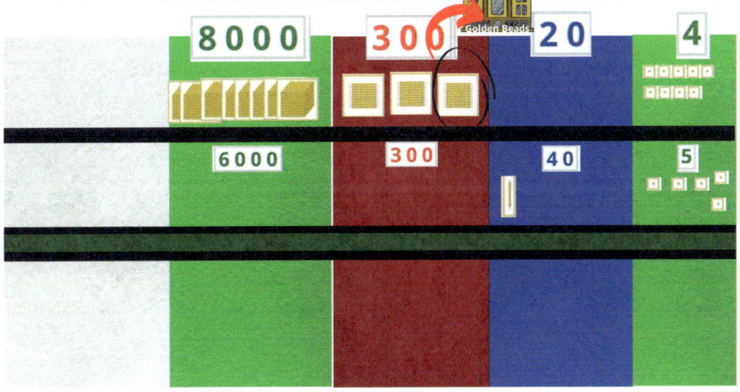

4

Now move down the ten-bars. If there are not enough, say, "There is only one ten. I need four tens. I am going to take one hundred from here and exchange it for ten ten-bars using the Golden Bead Bank." Place the tens in the top column containing the tens.
Then, re-count the beads you have already placed in the middle section, and take down the remaining needed tens to make forty.

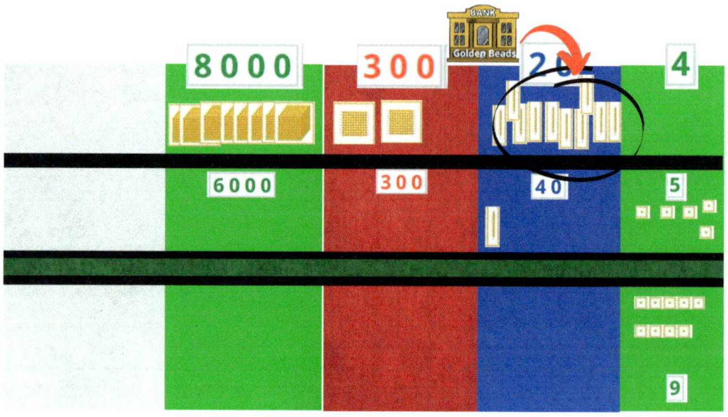

5

Place the tens in the top column containing the tens.
Then, re-count the beads you have already placed in the middle section, and take down the remaining needed tens to make forty.

6

(i) Then, re-count the beads you have already placed in the middle section, and take down the remaining needed tens to make forty.
(ii) Move the remaining ten-bar beads from the top section to the bottom section.
(iii) Retrieve the matching small numeral card.
(iv) Move on to the hundreds. Move the available hundreds down to the second row.
(v) Say, "There are only two hundreds. I need three hundreds. I will exchange one thousand cube for ten hundred squares using the Golden Beads Bank." Place the cube in the bank and take out ten hundreds.

7

Place the ten hundred squares in the top section of the hundred column.

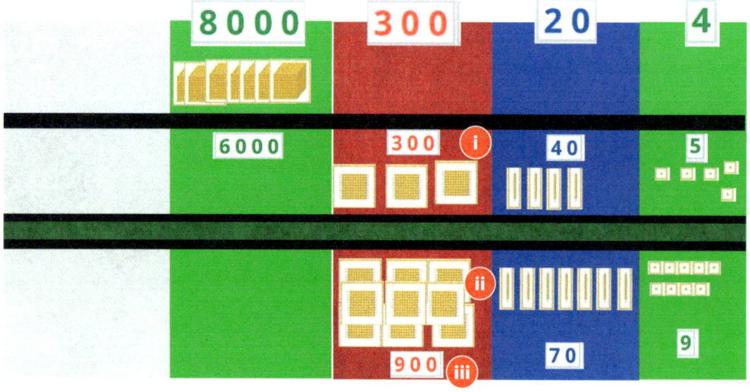

8

(i) Count the hundred squares you have already placed in the middle section, and move one more in place to make 300.
(ii) Move the remaining hundred squares to the bottom hundreds section of the mat to show the remainder.
(iii) Place the matching small numeral card.

10

Line the numeral cards all back up on the left side (*not shown) and copy the answers onto the problem grid, using the corresponding colors of colored pencils.

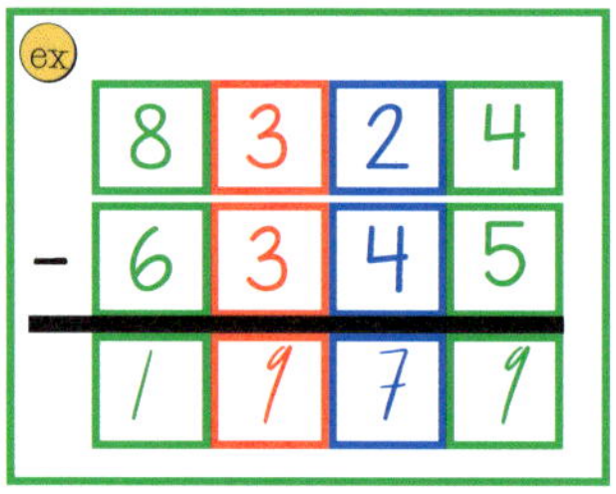

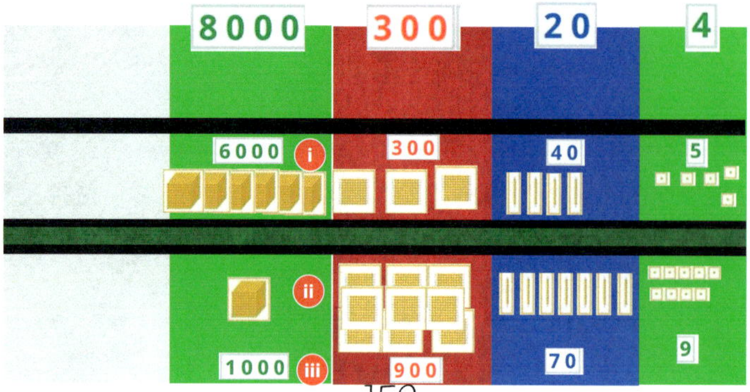

9

(i) Move down six thousand cubes into the subtrahend middle section of the thousands column.
(ii) Move down the remaining thousand cube into the bottom section to show the remainder.
(iii) Retrieve the matching small numeral card.

Golden Beads Dynamic Subtraction Problems

1

	4	9	5	7
-	2	7	5	9

2

	7	5	4	8
-	4	3	3	9

3

	8	8	0	3
-	1	3	6	7

4

	4	3	2	3
-	3	4	0	5

5

	9	7	5	6
-	7	4	6	8

6

	5	9	0	4
-	3	0	9	4

Golden Beads Dynamic Subtraction Problems

8

	4	9	4	7
-	3	9	6	3

9

	8	5	6	6
-	4	7	5	8

9

	7	9	2	0
-	2	8	4	3

10

	5	9	2	1
-	1	6	4	6

11

	6	7	1	3
-	5	8	8	0

12

	2	7	2	4
-		8	7	4

Golden Beads Subtraction: Cross Out the Subtrahend

Put an "X" over all of the beads you are subtracting. Count the beads that are left, and write the remainder using the proper colored pencils in the boxes on the right. Remember to start with the units!

Subtract 1245.

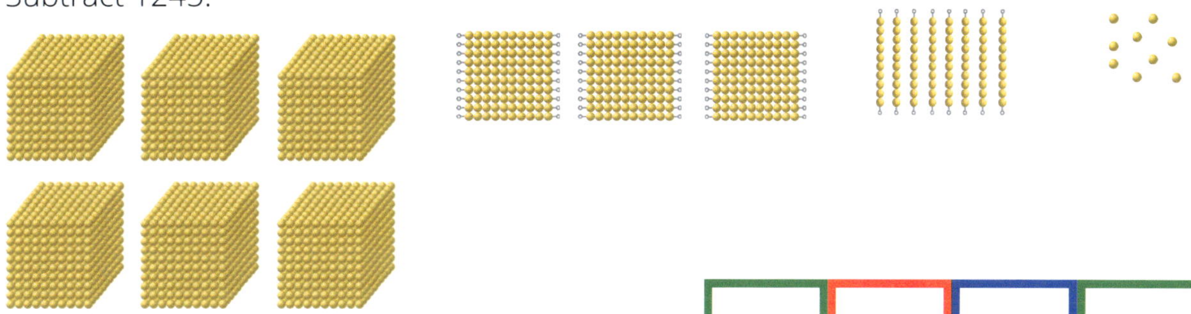

Subtract 4322.

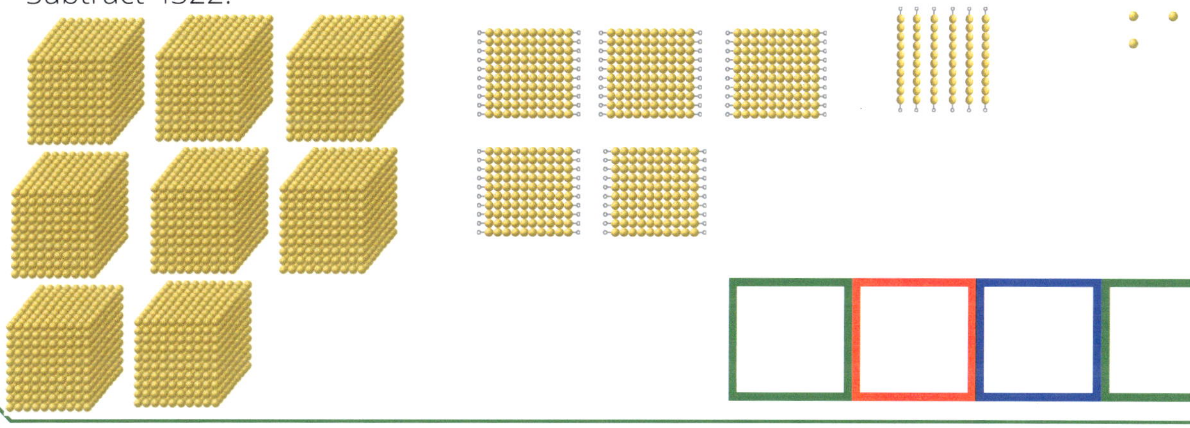

Subtract 2513.

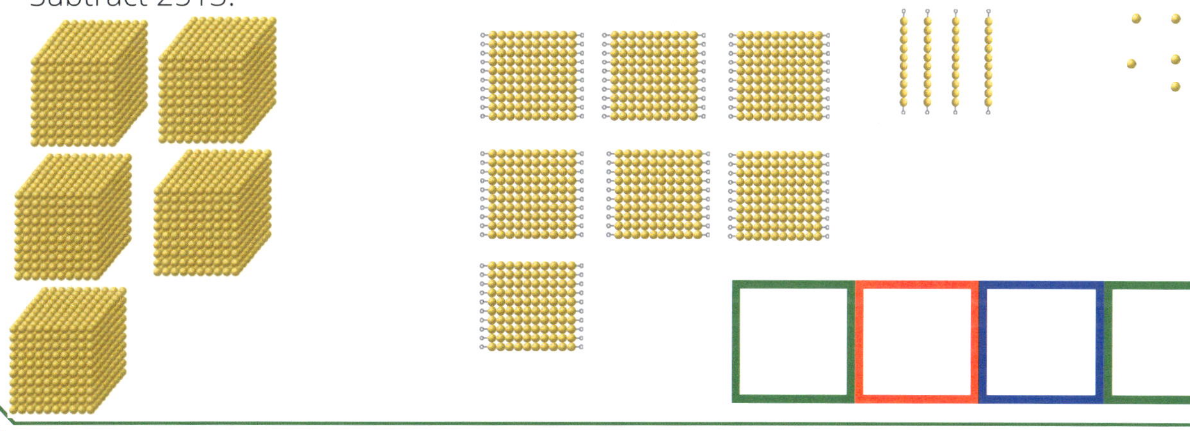

Golden Beads Subtraction: How Many Did I Take Away?

In each of the colored boxes, write the number of each amount of beads that was taken away to produce the remainder. Remember to start with the units!

Tip: It helps to put an "X" through each of the amounts in the minuend that also appear in the remainder. The beads that are not crossed out will equal the subtrahend.

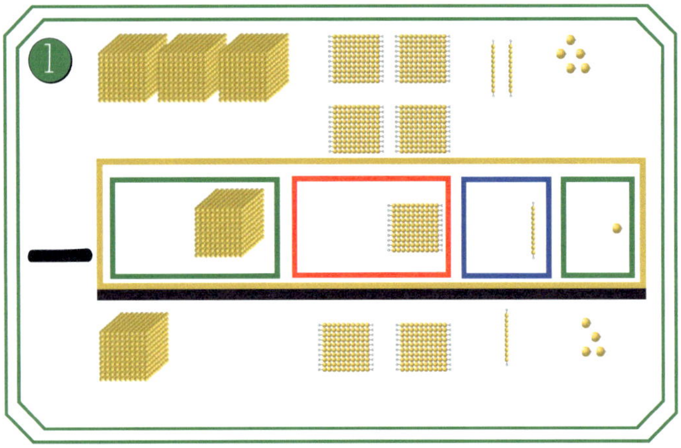

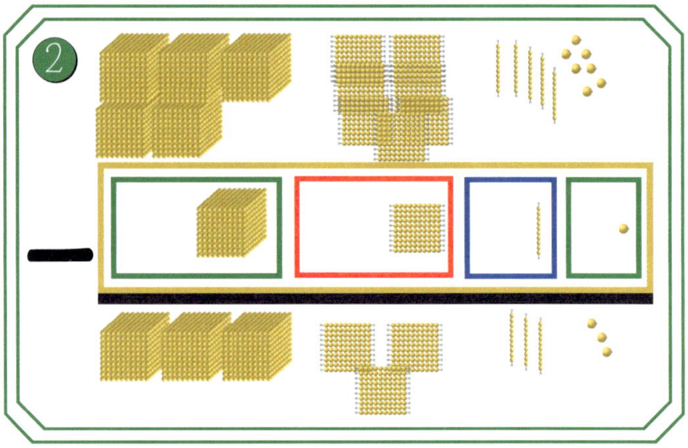

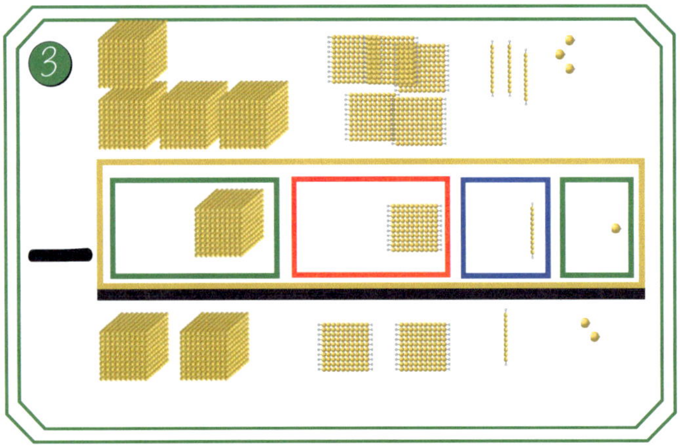

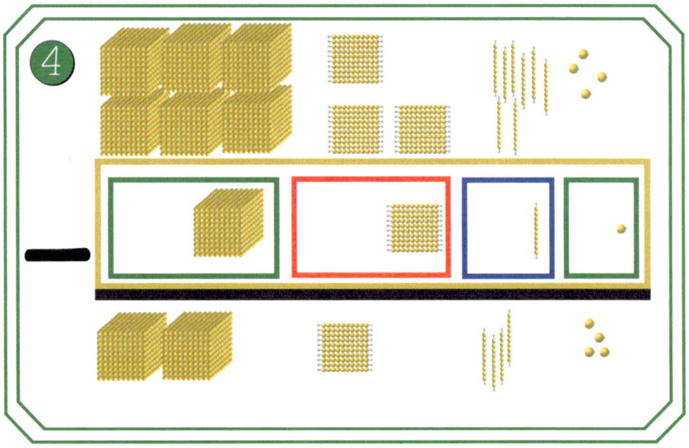

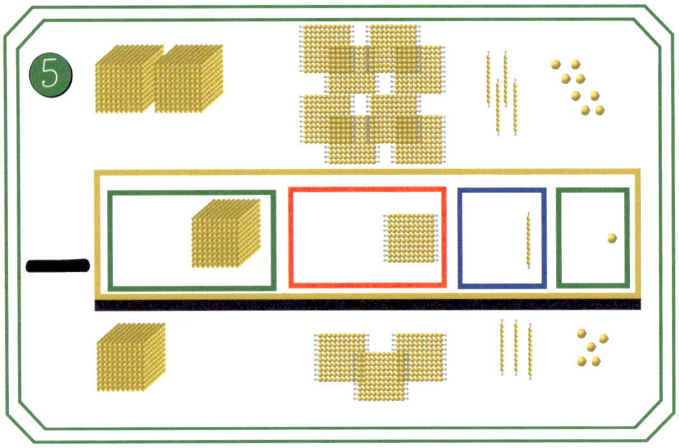

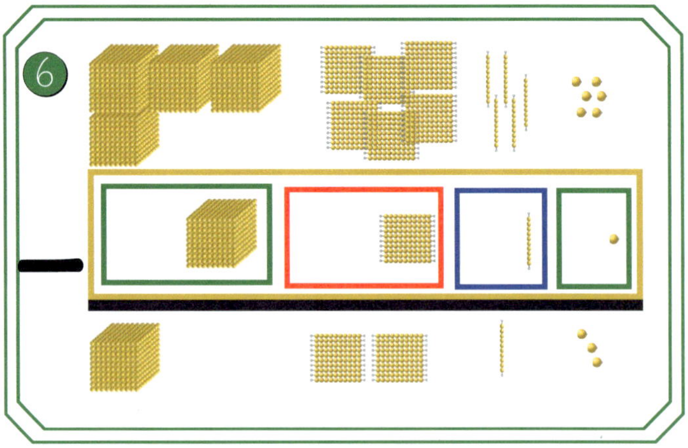

Golden Beads Multiplication Mat

Remove the mat from the workbook to use alongside the multiplication problems. Laminate or place in page protectors for durability.

Tape the middle of the mat together.

*Note: If you are using real Golden Beads or the larger Golden Beads Printables, you may want to replicate this mat in a larger size with felt or colored paper

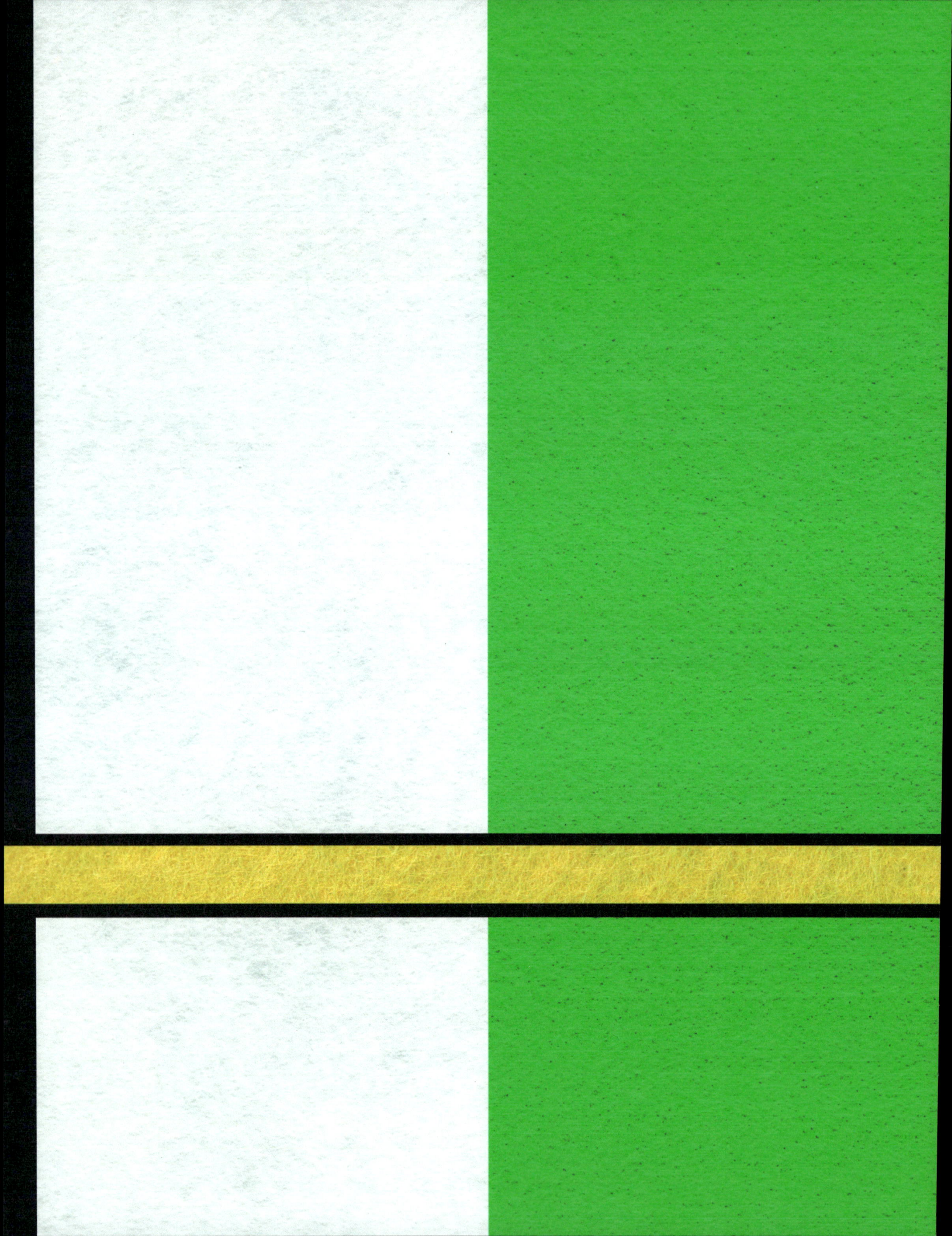

Golden Beads Static Multiplication

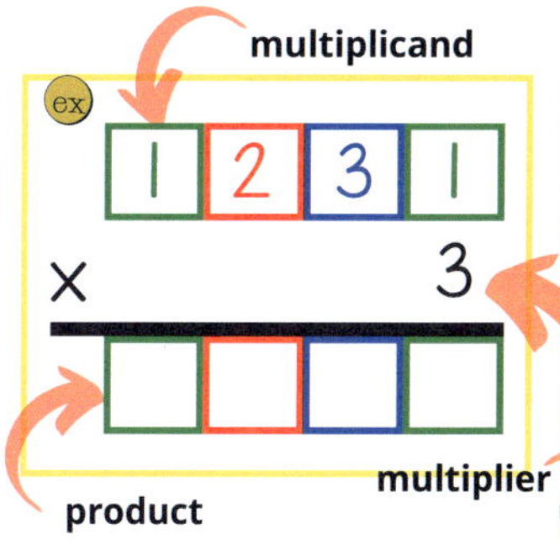

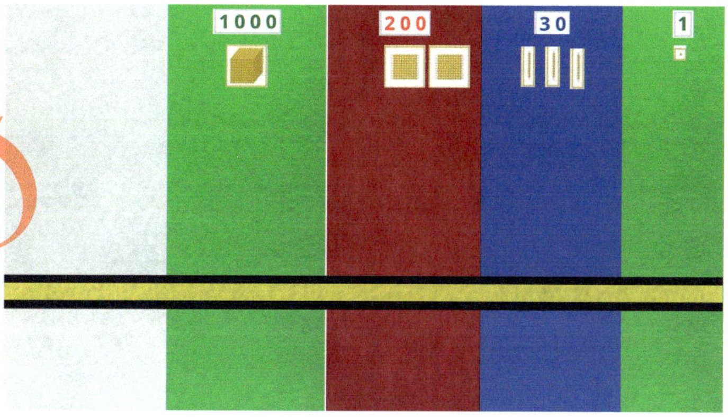

1

Point to the "x" symbol. Say, "This says, times. It means we are going to take this top number and add it to itself this number of times (point to the 3.) This is called **multiplication**." "This is the **multiplicand** (point to 1231). This is the **multiplier** (point to 3.)"

Invite your child to get the first set of small numeral cards and lay them out on the top left section of the mat, saying their names as you lay them out. ("One thousand, two hundred, thirty one.") Then separate them out and say each number **individually**. ("One thousand, two hundreds, three tens, and one unit.")

Retrieve the Golden Beads for each category, and say, "We have taken 1,231 one time."

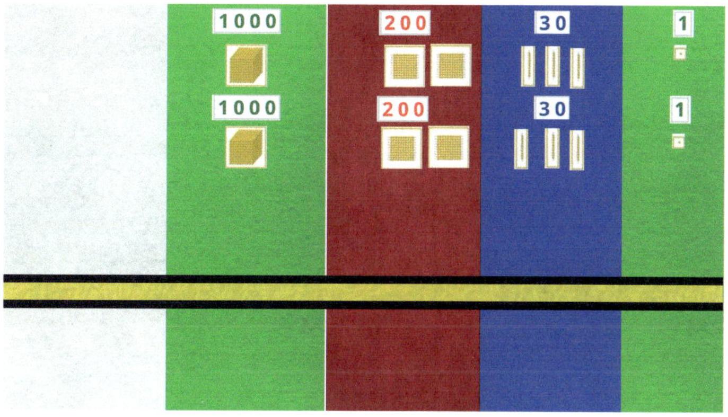

2

Point to the problem and say, "The multiplier **3** tells us we need to take the multiplicand **three times**. Let's fetch another set of small numeral cards for the multiplicand 1231." Fetch the cards, reading the numbers as you place them. Fetch the Golden Beads, and place them on the mat. When you are finished, say, "We have taken the multiplicand **two times**. We need to do it one more time."

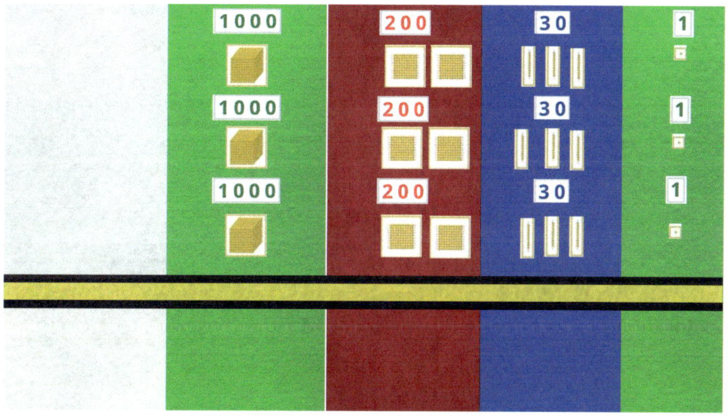

3

Retrieve a third set of small numeral cards for 1231, naming the number both ways as you first place it on the left side of the mat and then separating it out. Retrieve the corresponding Golden Beads.

4

Verify you have taken the number three times. Say, "We have taken the multiplicand **three** times" (point to "3" in the problem, then count 1-2-3 as you point to each row of numbers.)

"Now we can find the product." Move down the units to the bottom row. Count and retrieve the matching **large numeral card**. Continue for tens, hundreds, and thousands.

Finally, combine the number by stacking them (not shown) and read the final answer. Say, "One thousand two hundred thirty one times three is three thousand six hundred ninety three." Point to each line in the problem and say, "Multiplicand. Multiplier. Product. The product is our answer."

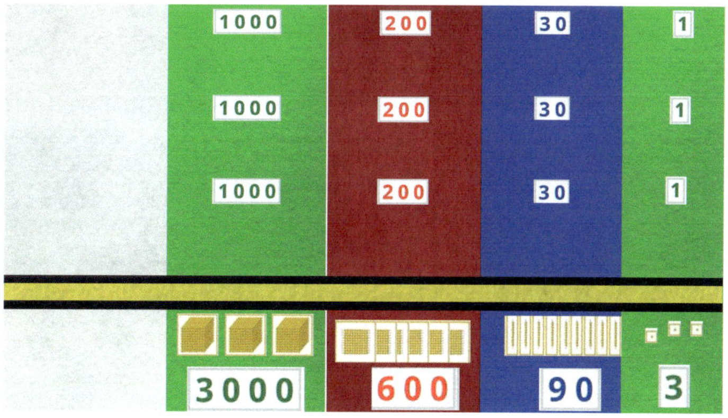

5

Copy the answers onto the problem grid, using the corresponding colors of colored pencils.

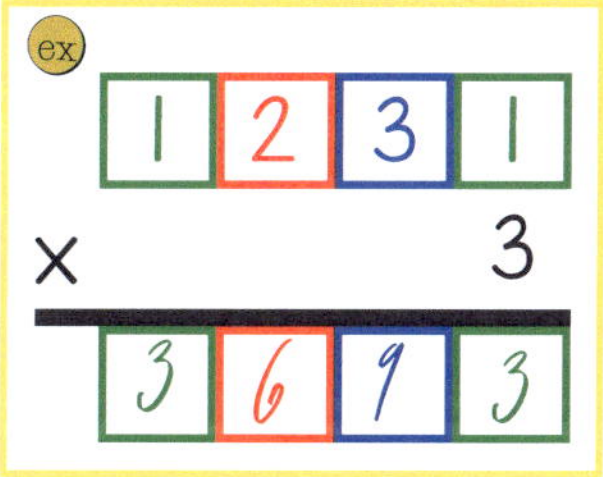

Golden Beads Static Multiplication Problems

1

2	1	3	4

× 2

2

3	2	1	2

× 3

3

4	0	2	2

× 2

4

2	2	0	3

× 3

5

3	2	1	0

× 2

6

1	2	3	3

× 3

Golden Beads Dynamic Multiplication

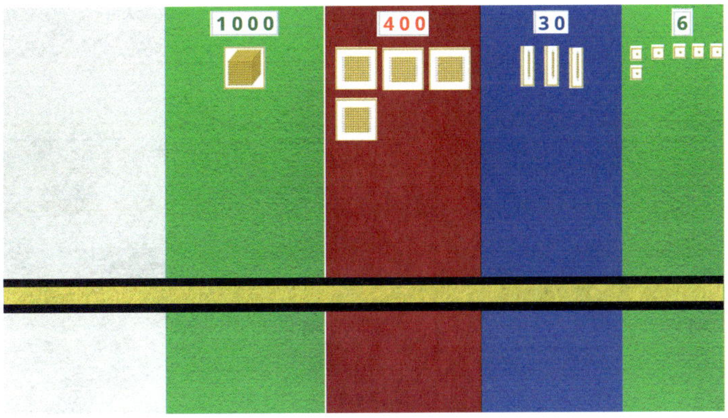

1

Point to the "x" symbol. Say, "This says, times. It means we are going to take this top number and add it to itself this number of times (point to the 2.) This is called **multiplication**." "This is the **multiplicand** (point to 1436). This is the **multiplier** (point to 2.)"
Invite your child to get the first set of small numeral cards and lay them out on the top left section of the mat, saying their names as you lay them out. ("One thousand, four hundred, thirty six.") Then separate them out and say each number **individually**. ("One thousand, four hundreds, three tens, and six units.")
Retrieve the Golden Beads for each category, and say, "We have taken 1,436 one time."

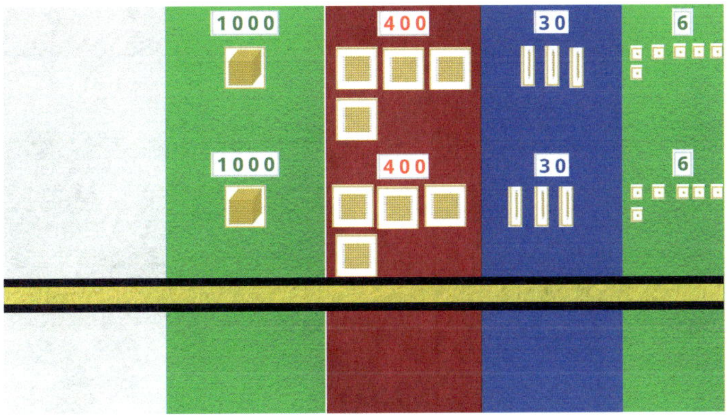

2

Point to the problem and say, "The multiplier **2** tells us we need to take the multiplicand **two times**. Let's fetch another set of small numeral cards for the multiplicand 1436." Fetch the cards, reading the numbers as you place them. Fetch the Golden Beads, and place them on the mat. When you are finished, say, "We have taken the multiplicand **two times**. Now we can find the product"

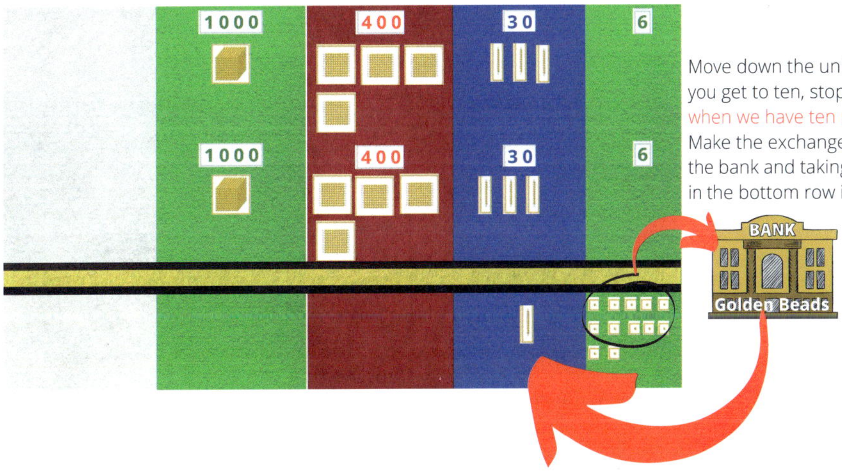

3

Move down the units to the bottom row. When you get to ten, stop and ask, "What can we do when we have ten units?" [Exchange for a ten-bar!] Make the exchange by placing the ten units into the bank and taking one ten-bar out and placing it in the bottom row in the tens place.

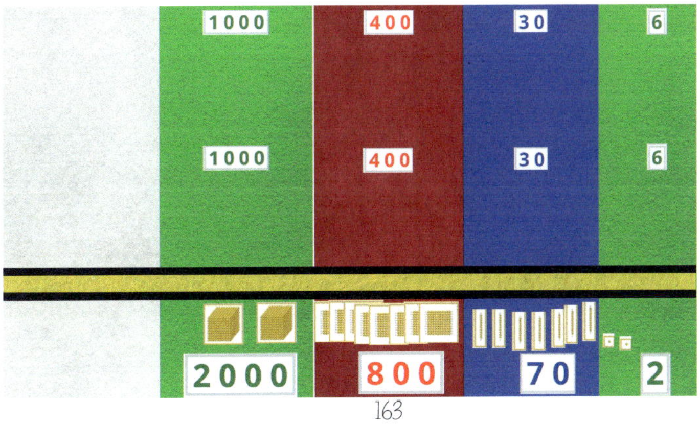

4

Count the remaining units after the exchange is made. [2] Fetch the large numeral card and place it below the units. Continue by moving down all the tens, counting the total, and making exchanges as needed.

*If you ever get zero in any category, do not place a numeral card.

5

Stack the large numeral cards and place them to the left. Read the problem again, pointing to the **multiplicand, multiplier, and product as you read the problem aloud.**
Copy the answers onto the problem grid, using the corresponding colors of colored pencils.

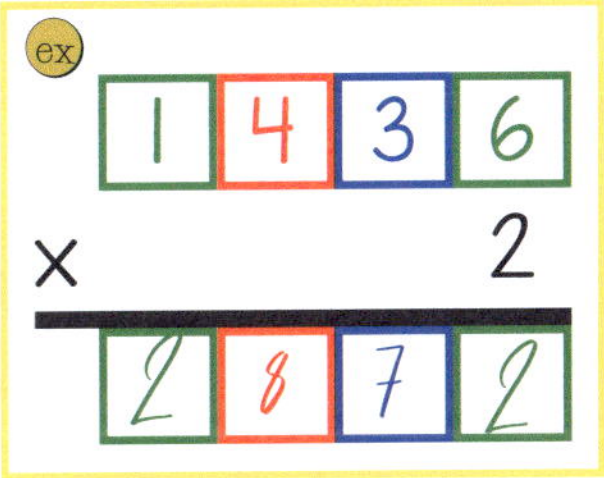

Golden Beads Dynamic Multiplication Problems

1.

4	3	7	4
×			2

2.

2	4	3	2
×			3

3.

1	7	2	8
×			2

4.

2	2	6	5
×			3

5.

2	9	1	5
×			2

6.

1	7	5	4
×			2

Golden Beads Dynamic Multiplication Problems

7

3	6	2	0
×			2

8

1	5	6	6
×			3

9

4	2	6	9
×			2

10

2	5	6	4
×			3

11

4	3	0	0
×			2

12

3	8	3	9
×			2

Golden Beads Multiplication: What is the Multiplier?

Show your child the parts of a multiplication problem again as a review. Ask your child: "What is the multiplier?" Invite your child to write the multiplier on the black line. The goal of this exercise is to reinforce the definition of a multiplier.

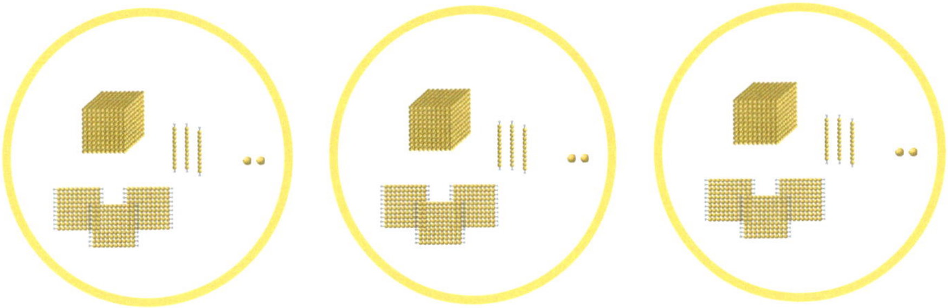

× ______

3297 3297 × ______

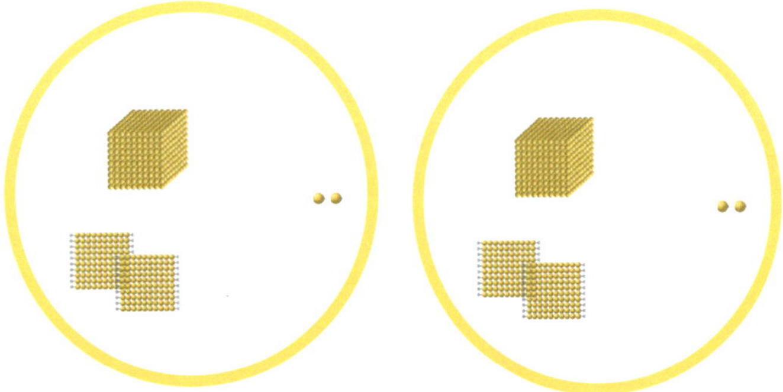

× ______

5149 5149 5149 × ______

How to Draw a Cube

Learn how to draw a cube so that you can draw a cube in later exercises to represent a one thousand cube of Golden Beads.

Try each step in the space below.

1

Draw a square.

2

Draw another square a little up and over.

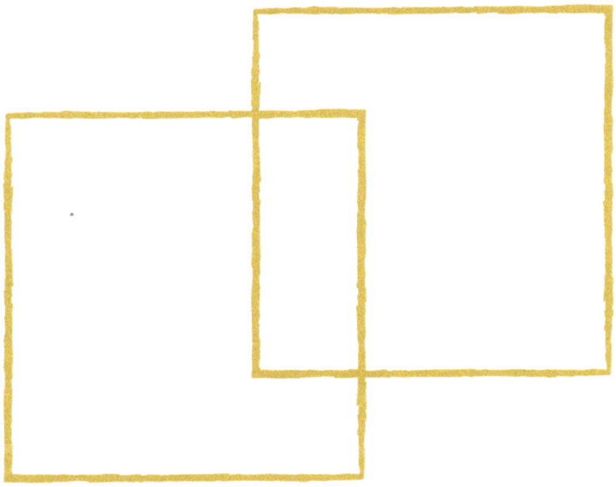

3

Connect the corners of the squares.

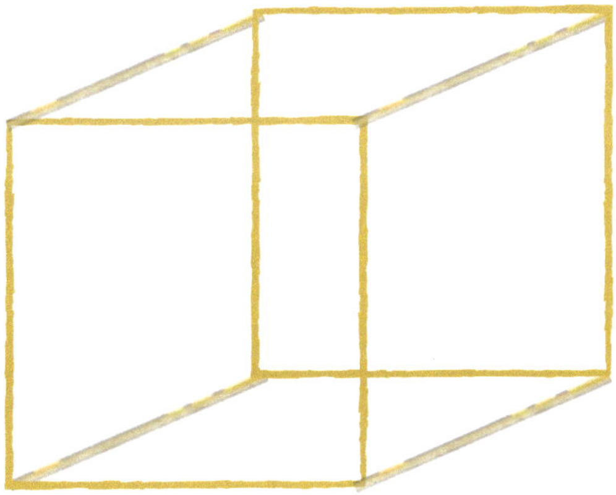

GO!

Now you can draw your own beads!

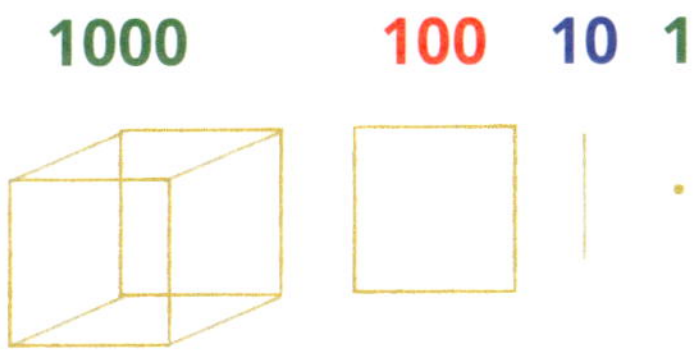

1000 100 10 1

Golden Beads Multiplication Activity

Below are two multiplication problems with the beads already set out.

1. Fill in the numerals of the multiplicand with the proper colors of pencil.
2. Add the Golden beads, starting with the units and working left, by drawing the beads in the space below the yellow line.
3. Write in the numerals of the product at the bottom of the box.
4. Write the problem in the "equation box" at the top right of the problem.
5. For the second problem, invite your child to also fill in the multiplier.

1000 100 10 1

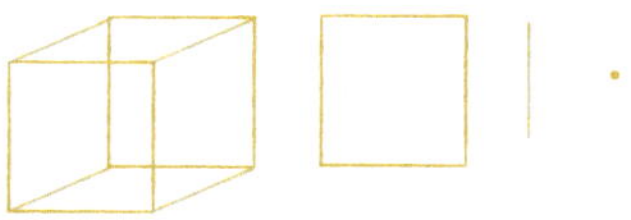

Tip: Cross out beads in the multiplicand after you have "moved" them by drawing them below the yellow line.

_000 _00 _0 _

_000 _00 _0 _

Equation Box

× 2

Draw the beads here for the product.

_000 _00 _0 _

_000 _00 _0 _

_000 _00 _0 _

_000 _00 _0 _

Equation Box

×

Draw the beads here for the product.

_000 _00 _0 _

Golden Bead Division Mats

Remove the four division mats from the workbook to use alongside the division problems. Laminate or place in page protectors cut to size and taped closed for durability.

Tape the middle of the mats together.

Separate the top larger mat from the three smaller mats.

*Note: If you are using real Golden Beads or the larger Golden Beads printables, you may want to replicate these mats with felt or colored paper.

Mat #1:

Mat #2:

Mat #3:

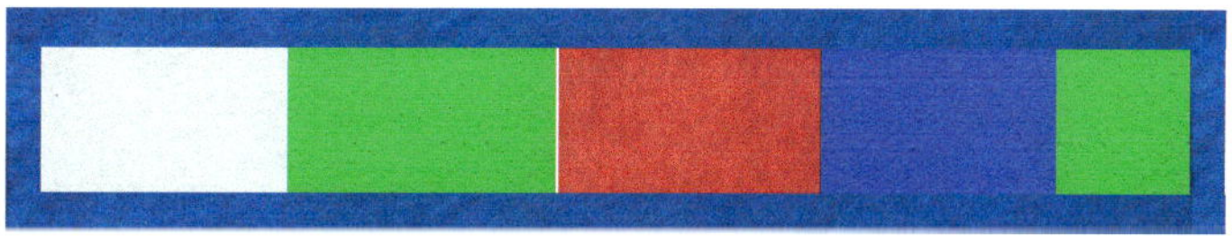

Mat #4:

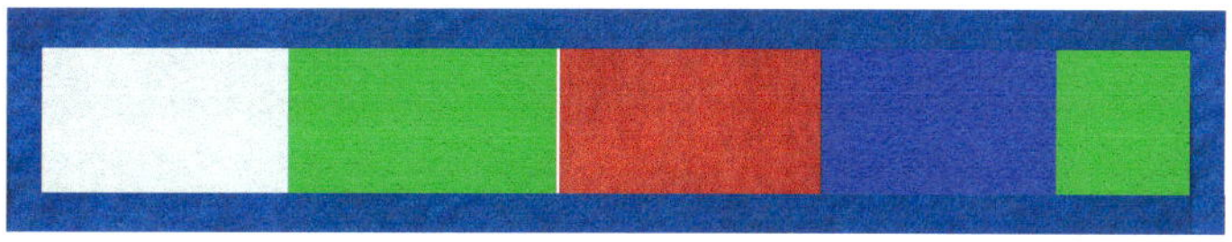

Golden Beads Static Division

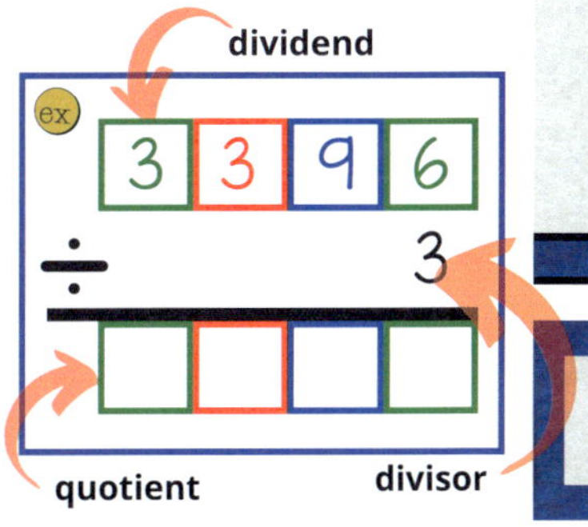

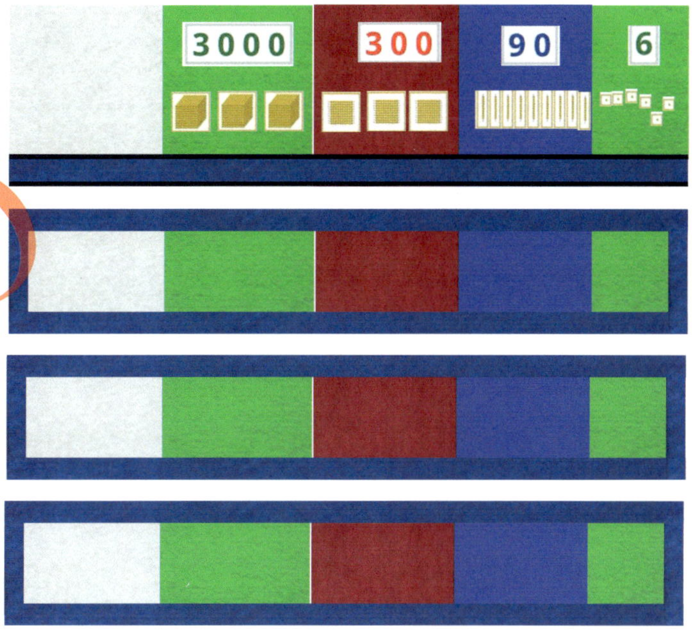

1

First, lay out the **large numeral cards** for **dividend** on the top, larger mat. Fetch the Golden Beads for each category.
Say, "We are going to **divide**." (Point to the division sign.) "This is a **division sign**. That means I am going to give each mat the same amount of beads. This number here (point to 3) is the **divisor**. This tells us how many mats we need." Count the smaller mats outlined in blue. Verify there are three mats.

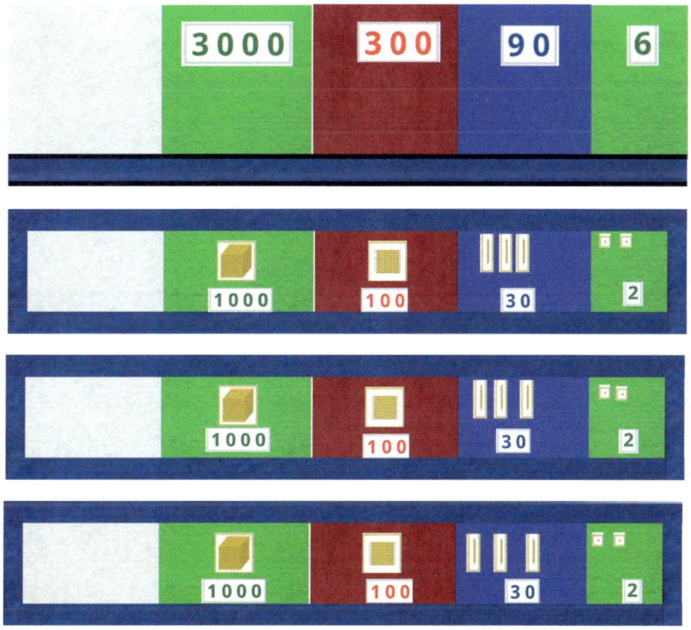

2

When dividing, always start with the **largest place value**. Set one cube at a time on the three mats, giving each mat an equal amount. Next, divide the hundreds, tens, and units equally, by placing one at a time on each mat.

3

Invite your child to fetch the small numeral cards for each mat.
Stack the numeral cards on the left-hand side of the mats. (*Not shown.)
Point out that each mat has the same numeral!

4

Read the problem, pointing to each number as you read. "Three thousand three hundred ninety-six is the **dividend**. We divided it by three, the **divisor**, to get the **quotient** one thousand one hundred thirty-three."
Invite your child to write in the number using the proper colored pencils onto the problem grid.

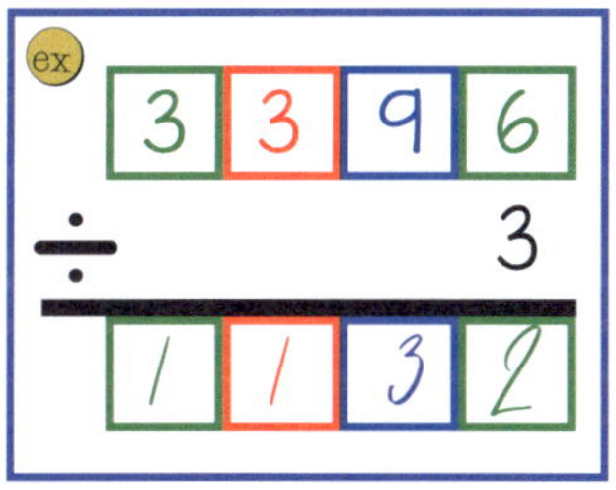

Golden Beads Static Division Problems

1

4	8	2	4
÷			2

2

6	3	3	9
÷			3

3

4	2	6	0
÷			2

4

9	3	0	6
÷			3

5

8	6	4	2
÷			2

6

6	0	3	9
÷			3

Exchanges

Draw a line from the quantity on the left to the equivalent quantity on the right.

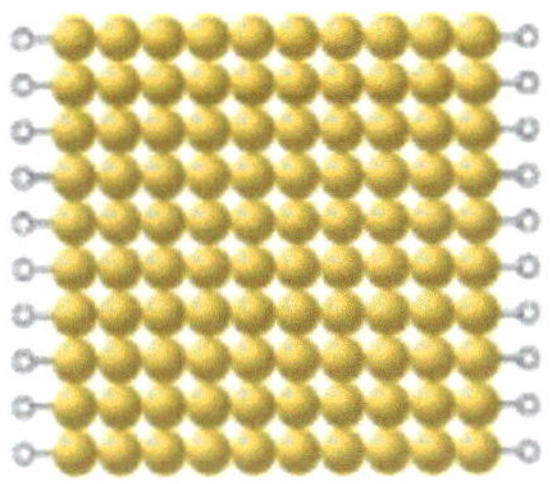

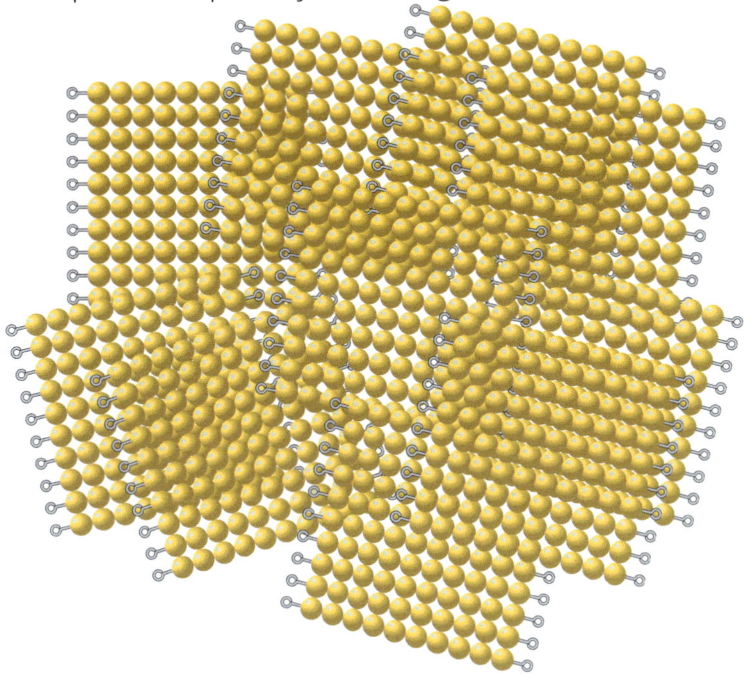

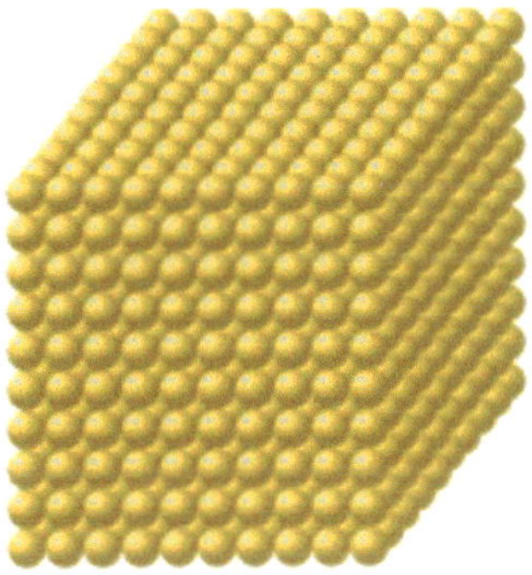

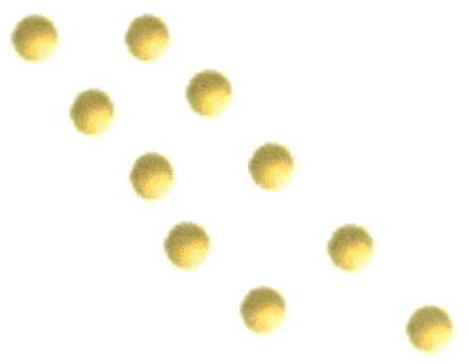

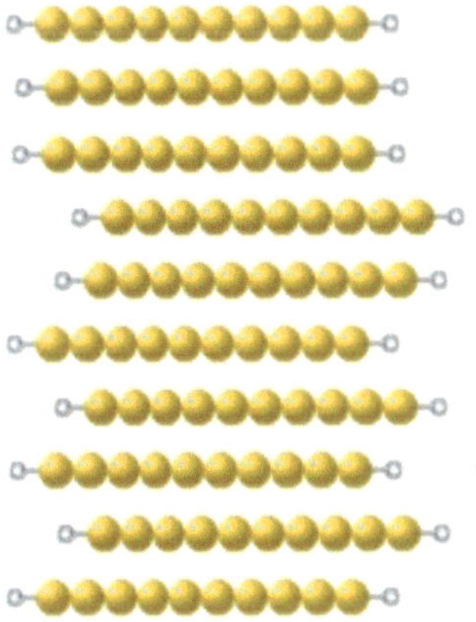

Golden Beads Dynamic Division

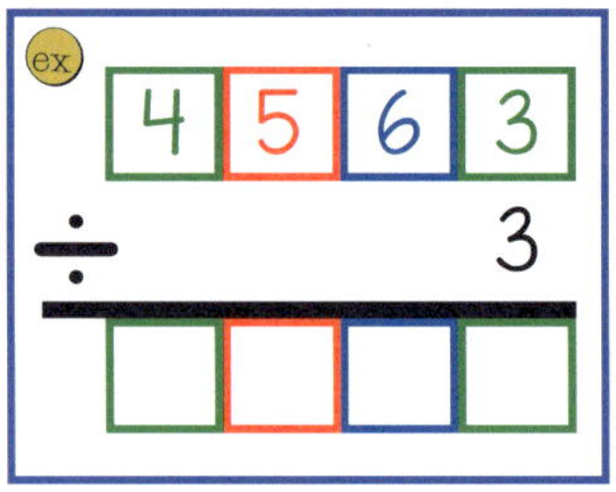

1

First, lay out the **large numeral cards** for **dividend** on the top, larger mat. Fetch the **Golden Beads** for each category.
Say, "We are going to **divide**." (Point to the division sign.) "This is a **division sign**. That means I am going to give each mat the same amount of beads. This number here (point to 3) is the **divisor**. This tells us how many mats we need." Count the smaller mats outlined in blue (not pictured in Step 1 on this page!). Verify you have three mats.

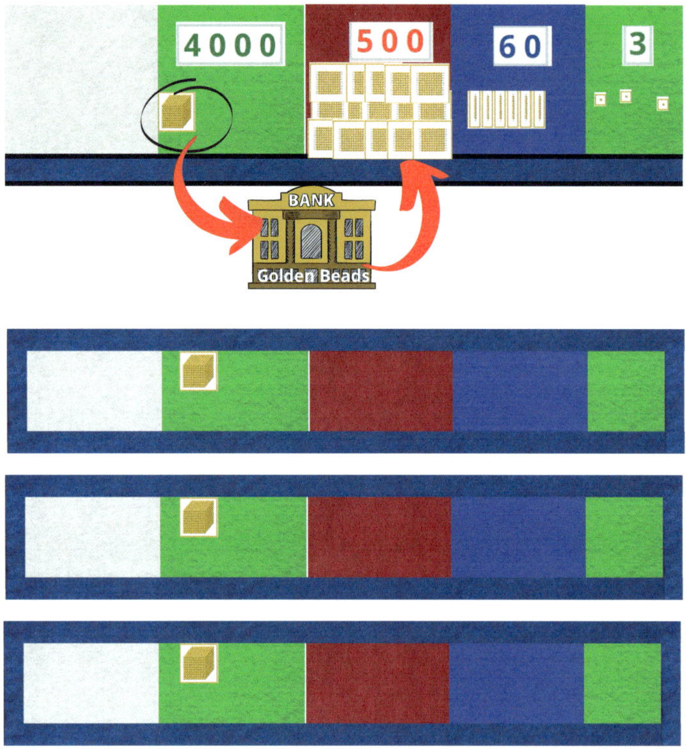

2

When dividing, always start with the **largest place value**. Set one cube at a time on the three mats, giving each mat an equal amount.
If you have any left over that you cannot split equally among the mats, say, "I cannot split this equally among the mats. I need to make an exchange for hundreds in the Golden Beads Bank." Make the exchange, ten of the smaller units for each one of the larger units. In this case, leave the one thousand cube in the bank and take out ten hundred squares. Place the ten hundred squares with the other hundred squares on the large mat.

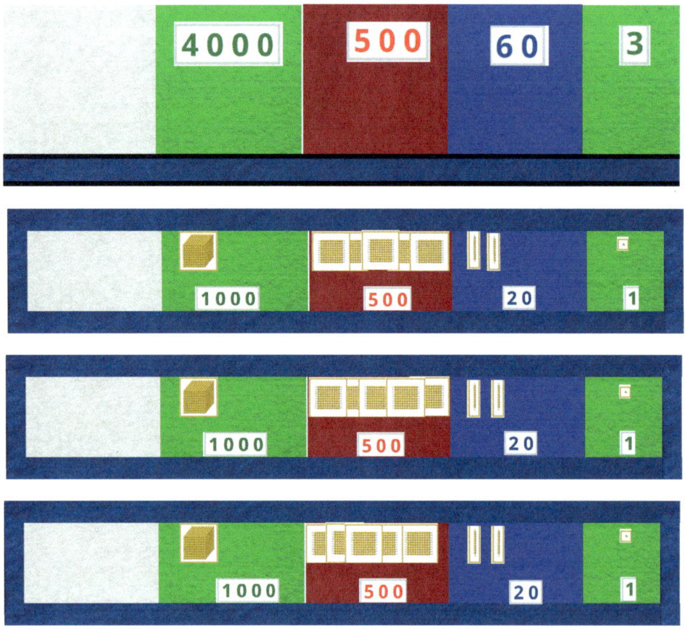

3

Proceed to divide the rest of the beads, making any exchanges when you reach one of one place value.
Invite your child to fetch the **small numeral cards** for each mat.
Stack the numeral cards on the left-hand side of the mats. (*Not shown.)
Point out that each mat has the same numeral!

4

Read the problem, pointing to each number as you read. "Four thousand Five hundred sixty-three is the **dividend**. We divided it by three, the **divisor**, to get the **quotient** one thousand five hundred twenty-one."
Invite your child to write in the number using the proper colored pencils onto the problem grid.

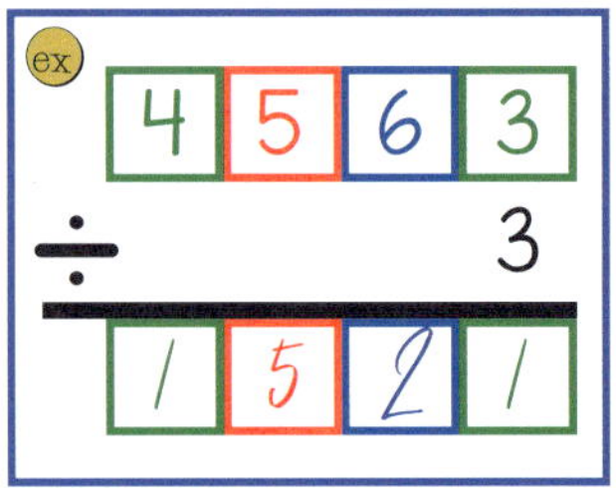

Golden Beads Dynamic Division Problems

1

5	6	4	6

$\div$ 2

2

6	7	2	9

$\div$ 3

3

4	2	9	8

$\div$ 2

4

7	5	4	2

$\div$ 3

5

5	3	1	2

$\div$ 2

6

6	8	4	9

$\div$ 3

Golden Beads Dynamic Division Problems

7

9	2	2	8

÷ 2

8

4	5	6	3

÷ 3

9

8	6	5	2

÷ 2

10

4	2	7	5

÷ 3

11

9	7	1	4

÷ 2

12

9	1	5	6

÷ 3

Golden Beads Division: Remainders

Sometimes when we divide, we have beads leftover. We call the leftover beads the "remainder." Each of the problems on this page has a remainder. Write the quotient, then write the remainder in the box.

1.

4	7	3	5

÷ 2

R = ___

2.

9	5	1	8

÷ 3

R = ___

3.

8	3	4	3

÷ 2

R = ___

4.

5	5	1	3

÷ 3

R = ___

5.

1	2	1	3

÷ 2

R = ___

6.

4	7	0	5

÷ 3

R = ___

Golden Beads Division: Zero in the Dividend

When there is a zero in the dividend, it acts as a placeholder. Zero divided by any number is zero. Work through these problems with the Golden Bead Division Mats to demonstrate how zero functions in the dividend. If there is a remainder, write it on the line. If there is no remainder, write R=0.

1. 5640 ÷ 2

R = ___

2. 6309 ÷ 3

R = ___

3. 4078 ÷ 2

R = ___

4. 7540 ÷ 3

R = ___

5. 5302 ÷ 2

R = ___

6. 5047 ÷ 3

R = ___

Golden Beads Division Activity

Can you divide the beads on the table equally among your three friends? Evenly distribute the beads on each tray, starting with the largest beads. Draw your beads like you have been shown before, using cubes, squares, lines, and dots. Cross out the beads on the table with an "X" after you have placed them on your friends' trays. BONUS: Write the quotient!

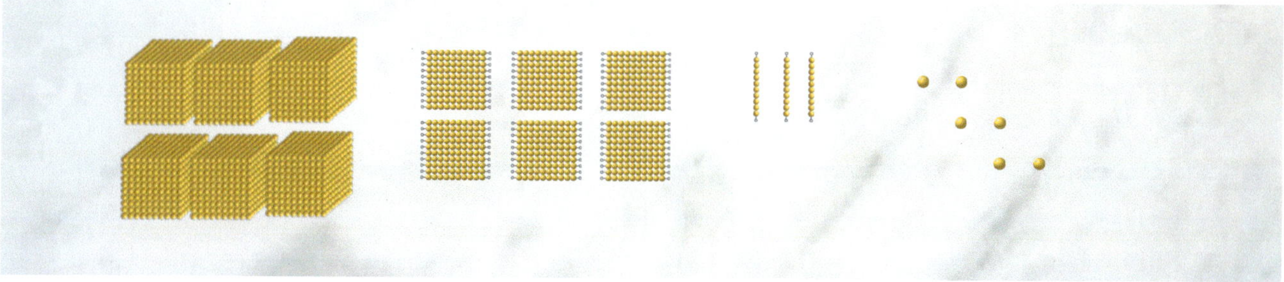

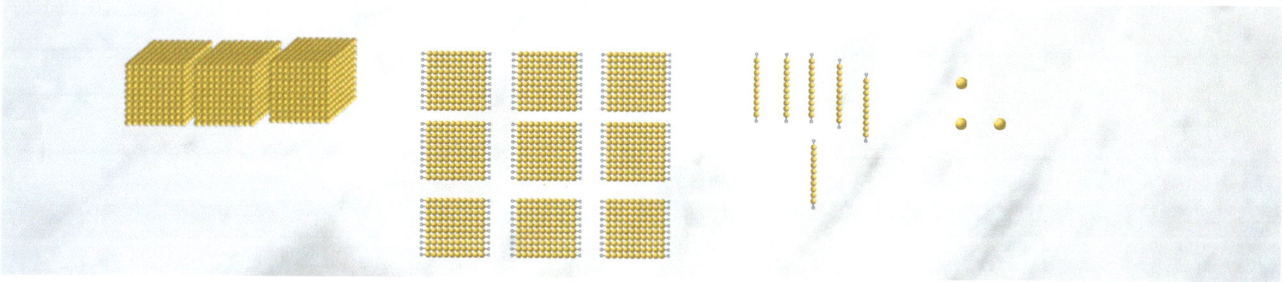

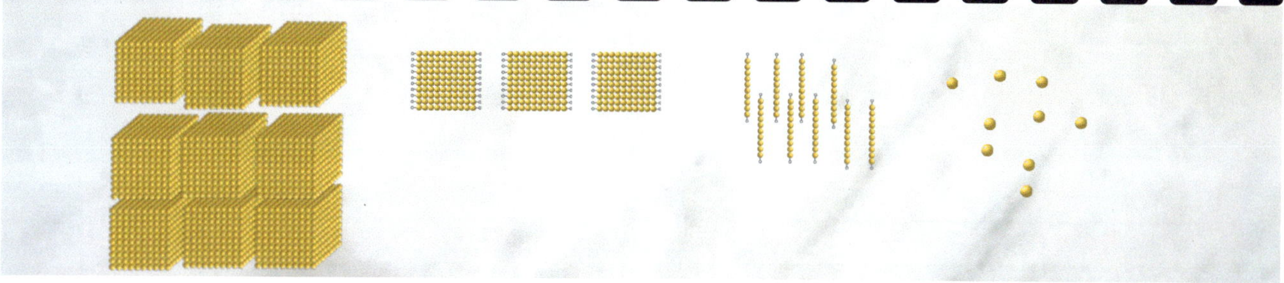

Operations Symbols

Color each symbol as follows: **addition - red**, **subtraction - green**, **multiplication - yellow**, **and division - blue.** Then trace and write the operation for each symbol.

add

subtract

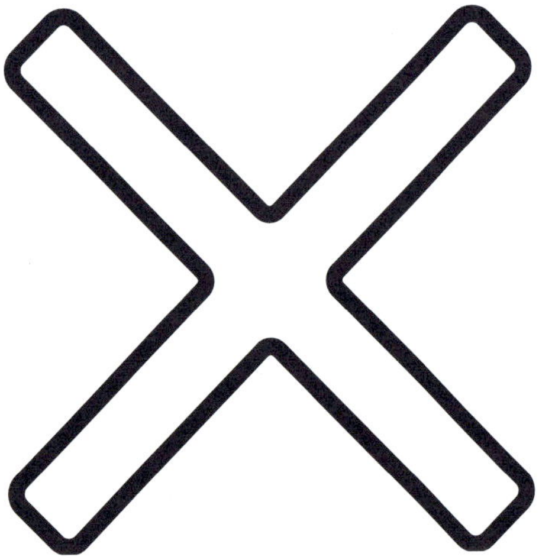

multiply

divide

Looping Review Exercises

As children progress through the Montessori math sequence, they will continually review concepts previously mastered as they work with new material. To keep those math muscles even stronger, you may consider inviting your child to go back and review prior lessons. Here you will find 60 days worth of "review" lessons that incorporate concepts from Book 1 and this book (Book 2.)

These exercises are set up to be completed one colored box at a time. To hone additional math skills, invite your child to shade in the day of the week bubble, write the date in the blank "tab" box, and fill in the time your child starts the activity and then the time your child ends the activity. Depending on the age, ability, and interest of your child, you may want to start off filling in the times for your child. When he/she shows interest, invite your child to take over the job!

None of these exercises require additional printed or physical materials. They do not require cutting and pasting. Your child can complete these exercises with just a pencil and crayons (or colored pencils).

Your child will review previous math skills and newly acquired math skills as he/she loops through the skills twice. These are intended to be completed alongside the new math work presented in this *Montessori Math Workbook - Primary Book 2.*

Some Skills that will be reviewed in these exercises are:

- Number Rods and Operations with Number Rods
- Short Bead Stair and Operations with Short Bead Stair
- Distinguishing between even and odd numbers
- Make Ten Snake Game
- Teen Number recognition and formation
- Counting by tens
- Counting 1-100
- Short Bead Chains
- Squares
- Golden Beads and Decimal System Introduction Review
- Golden Beads Operations

*Remember, if your child needs help writing, you can use these "badges" and be your child's scribe. I included an extra set on this page, but you can use the ones from the beginning of the workbook, as well. They are the same.

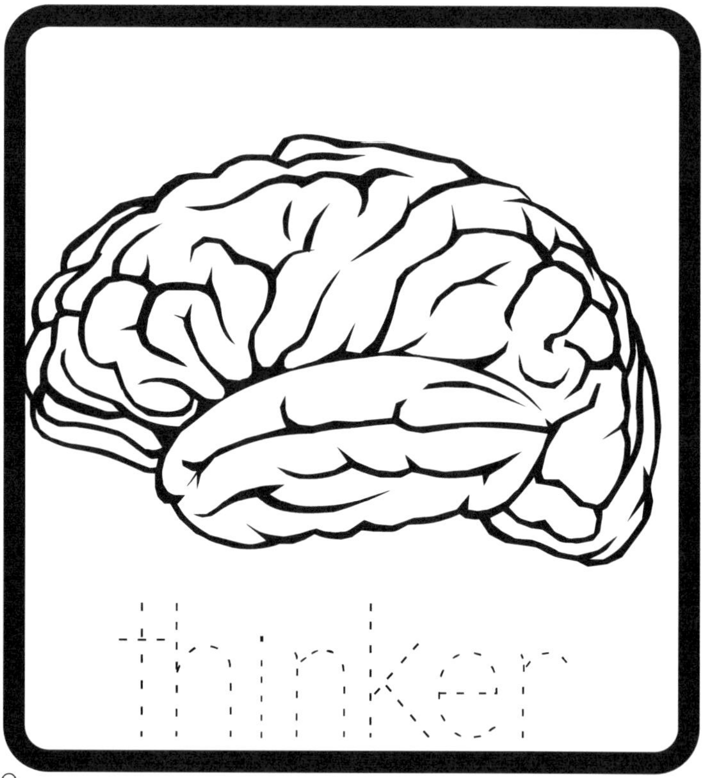

Looping Review Exercises

1

M T W Th F S Su

- This is the rod of __________.
- Underneath this rod, draw the rod that is one more than this rod.
- Above this rod, draw the rod that is two less than this rod.

2

M T W Th F S Su

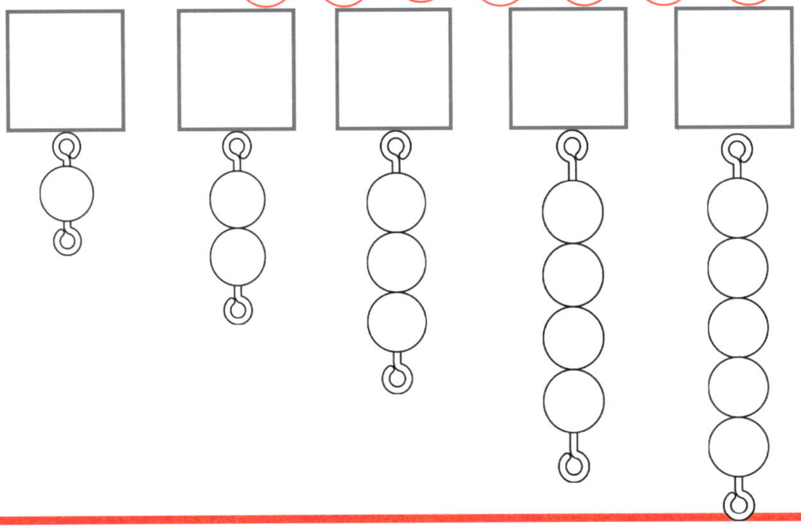

Color the bead bars their proper colors, then write the corresponding numeral in each box.

3

M T W Th F S Su

Write the number in the top left corner of each box. Circle "even" or "odd."

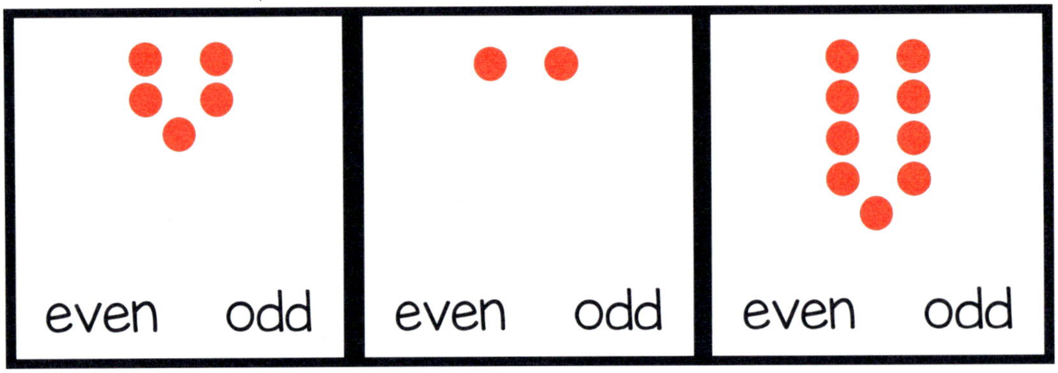

M T W Th F S Su

4

Write the numeral that represents the sum on the line.

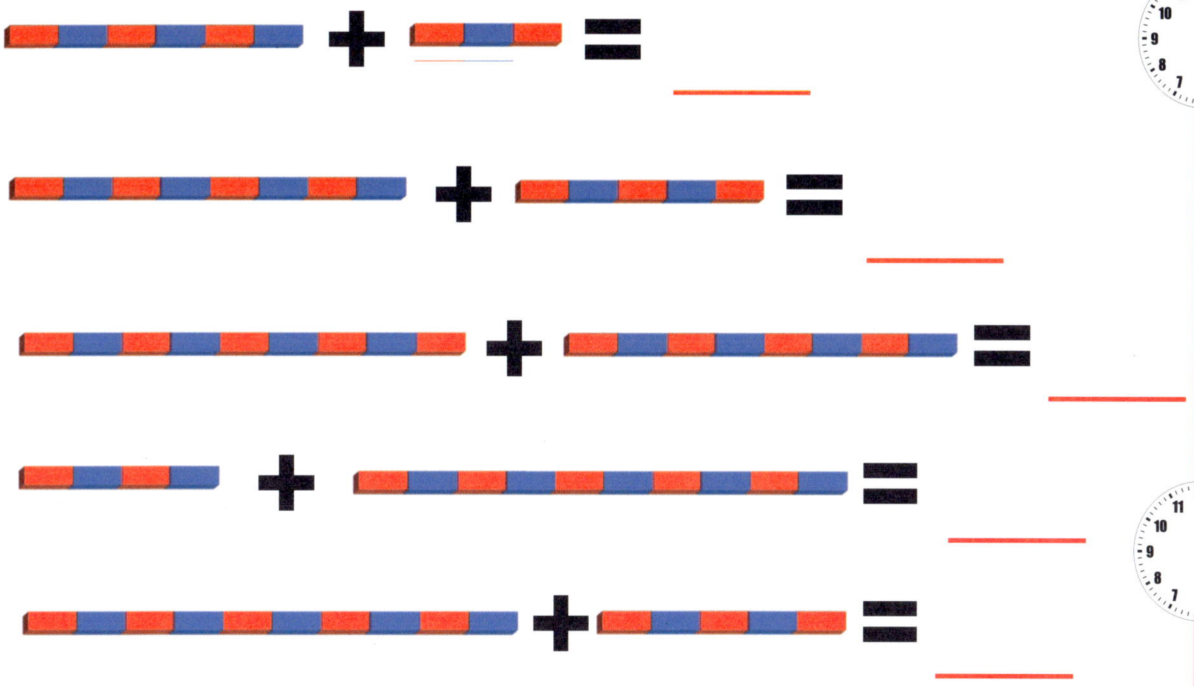

M T W Th F S Su

5

Write the correct symbol in the circle.

Greater than >

Less than <

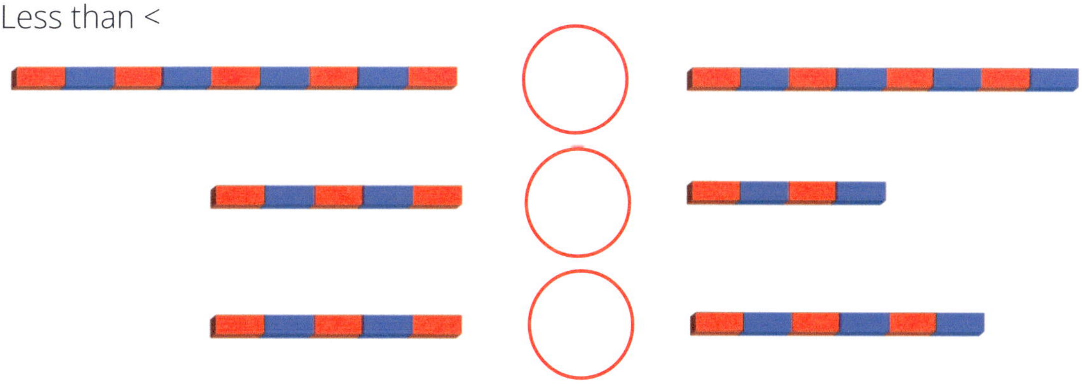

7 ◯ 9

8 ◯ 2

Hint: The alligator eats the greater number. We call him the "greater gator."

Looping Review Exercises

6

M T W Th F S Su

Draw the rod that you need to make ten. Then write the addition equation.

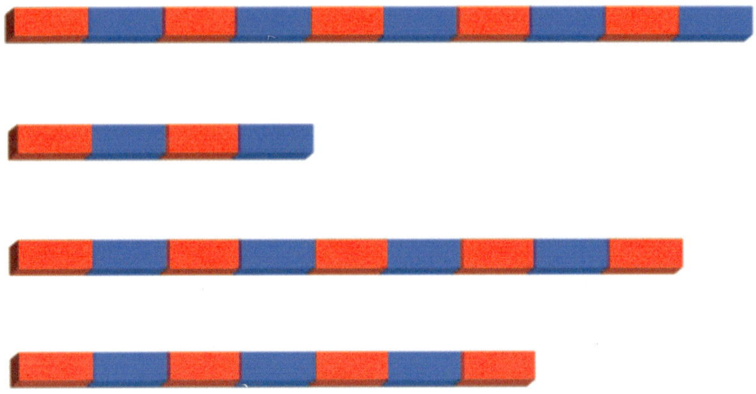

10

___ + ___ = ___

___ + ___ = ___

___ + ___ = ___

7

M T W Th F S Su

Draw the number rod or write the numeral representing the remainder.

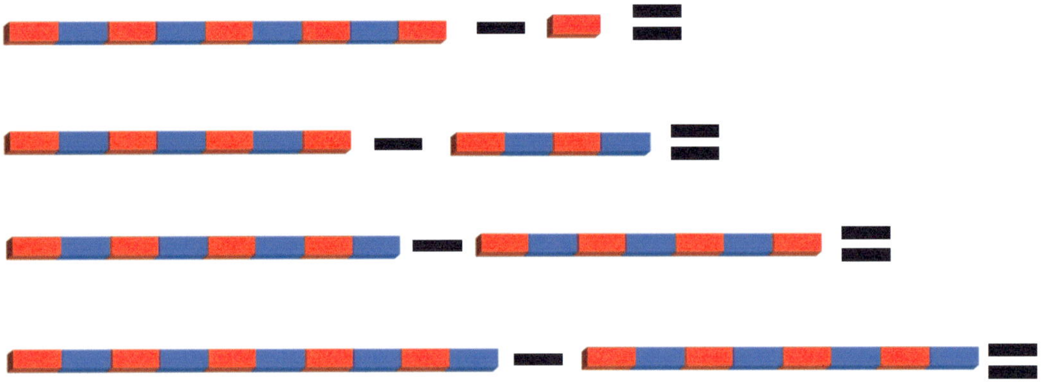

8

M T W Th F S Su

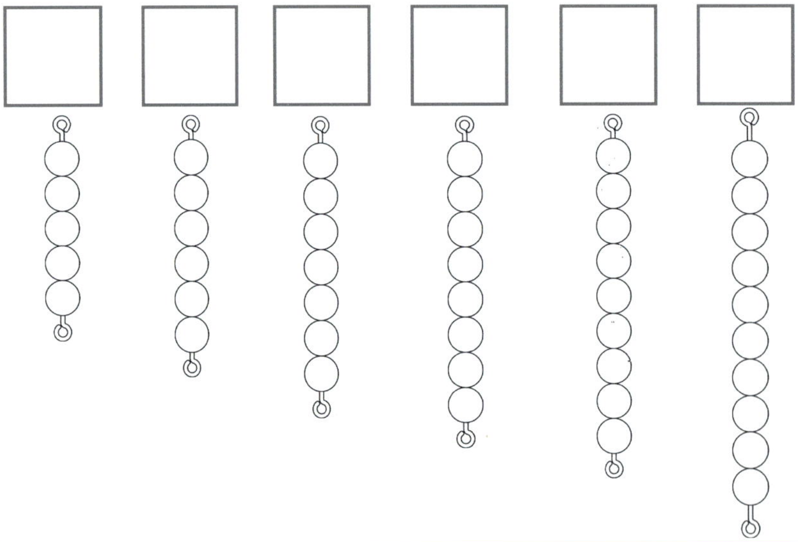

Color each bead bar its proper color. Then write the numeral in its corresponding box.

9

M T W Th F S Su

Using colored pencils, draw the bead bar you would need to add to make ten.

 + =10

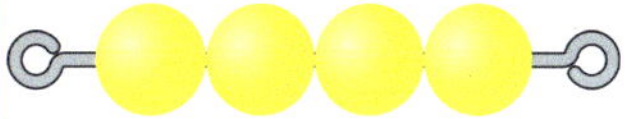 + =10

 + =10

 + =10

10

M T W Th F S Su

Make Ten Snake Game (without cutting and pasting)

Count to ten. Draw a pencil mark after the tenth bead. Cross out the ten beads you counted and draw a golden bead ten-bar to start your new snake. Keep counting from your pencil mark until you get to 10. Draw a pencil mark after the tenth bead. Cross out the ten beads you counted and draw another golden bead ten-bar on your new snake. If there are less than ten beads at the end, draw a black and white bead bar.

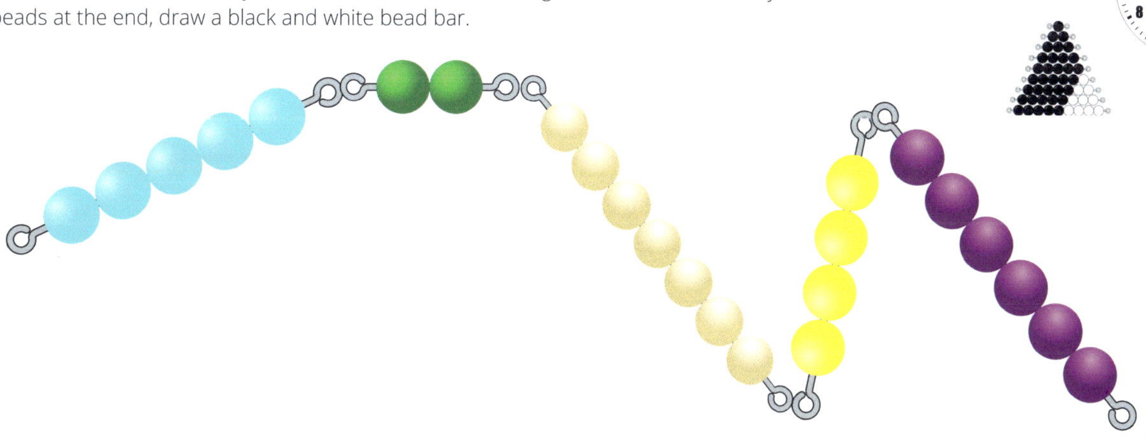

11

M T W Th F S Su

Color the Short Bead Bars, then write the teen number made.

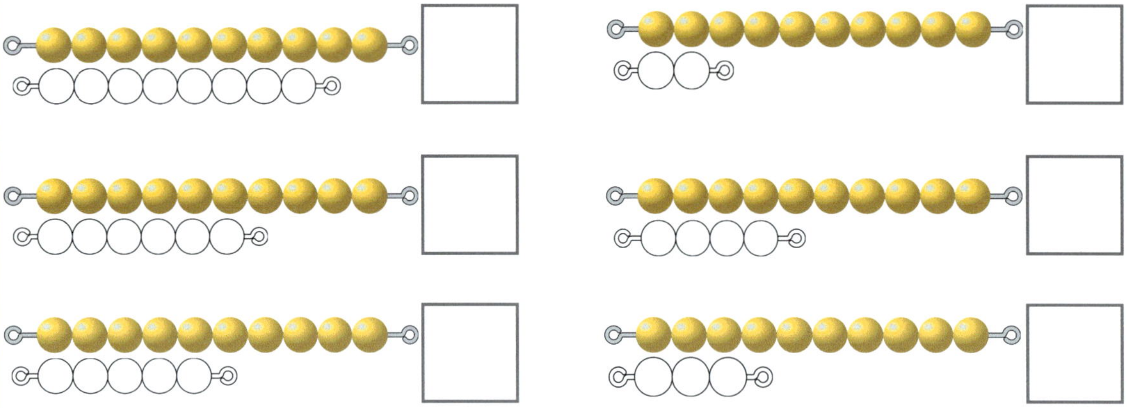

12

M T W Th F S Su

Draw a line from the numeral tile to the space it belongs on the teen board for numbers 11-15.

Then, write in the numerals needed to make 11-15 on the other board.

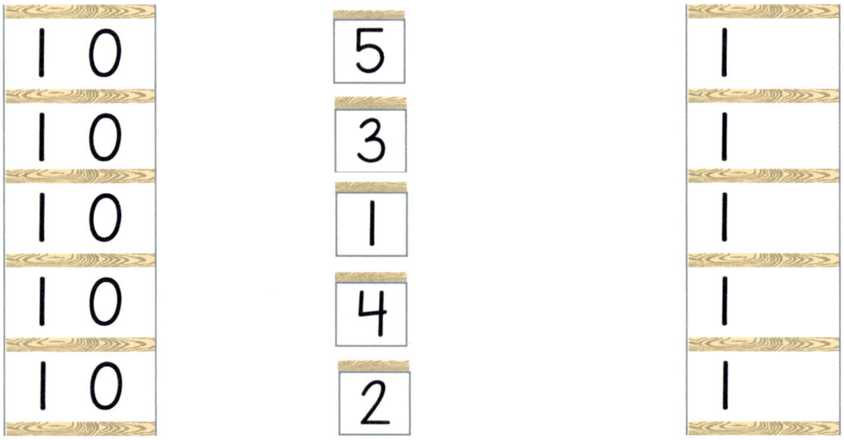

13

M T W Th F S Su

Write the numerals on the wood tiles along the top of the teen bead hanger.

Note: If the tiles are too small, invite your child to write them along the bottom of this page and draw a line to the tile.

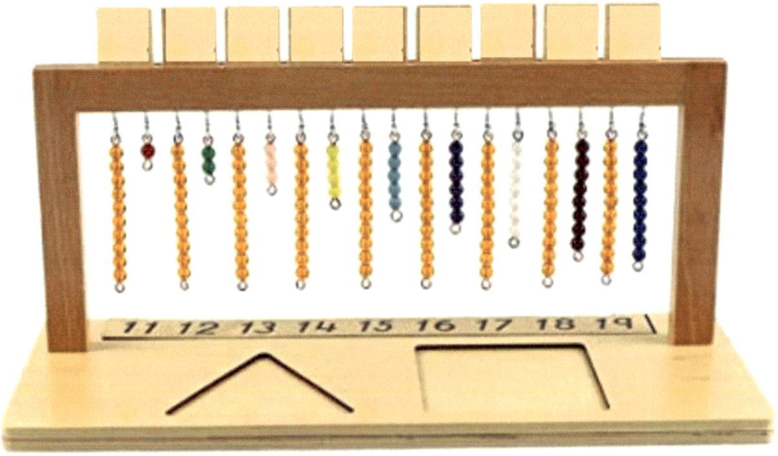

14

M T W Th F S Su

What teen value is made from combining the two number rods? Write the teen numeral in the box.

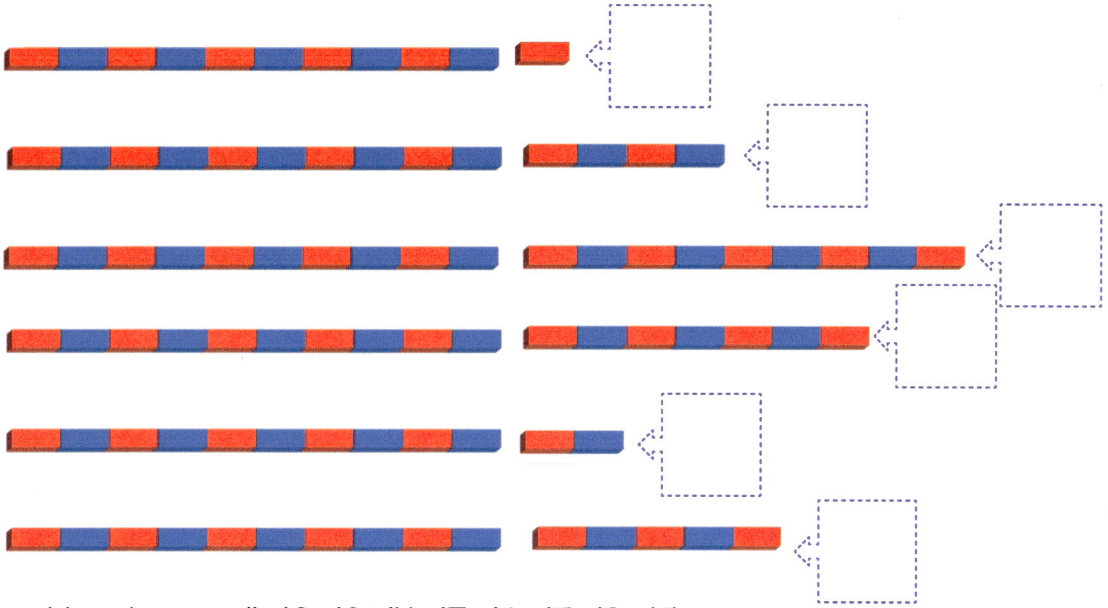

Teen Numbers: 11 12 13 14 15 16 17 18 19

15

M T W Th F S Su

How many tens are represented in each picture?
Write the numeral in the box. (Review numerals at the bottom of the page.)
(Two tens is written as 20.)

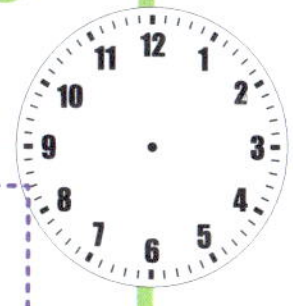

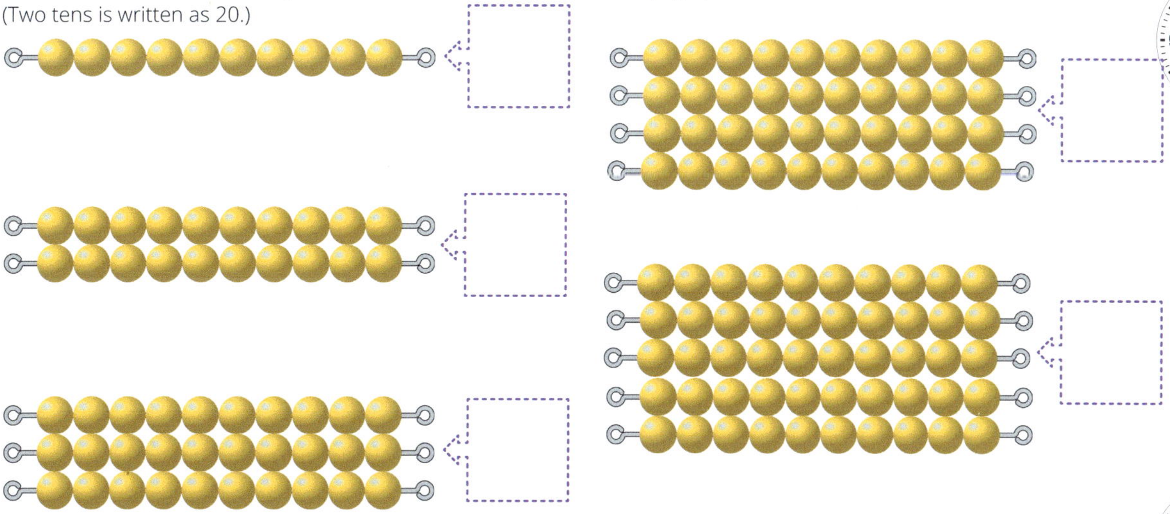

Tens: 10 20 30 40 50 60 70 80 90

Looping Review Exercises

16

M T W Th F S Su

Review the ten numerals with your child by first pointing and naming 10-90. Then check off the items your child completes. Go back to the ones missed.

- ◯ Show me twenty.
- ◯ Show me ninety.
- ◯ Show me seventy.
- ◯ Show me fifty.
- ◯ What is this? (Point to thirty.)
- ◯ What is this? (Point to sixty.)
- ◯ What is this? (Point to ten.)
- ◯ What is this? (Point to eighty.)
- ◯ What is this? (Point to forty.)

10	60
20	70
30	80
40	90
50	

17

M T W Th F S Su

10	60
20	74
30	80
40	90
50	

Draw the Golden Beads represented by the circled numeral. Use a line for ten-bars and dots for unit beads.

10 1

18

M T W Th F S Su

Color the beads along the 100 Chain according to the key below the chain.

Start counting here at 1.

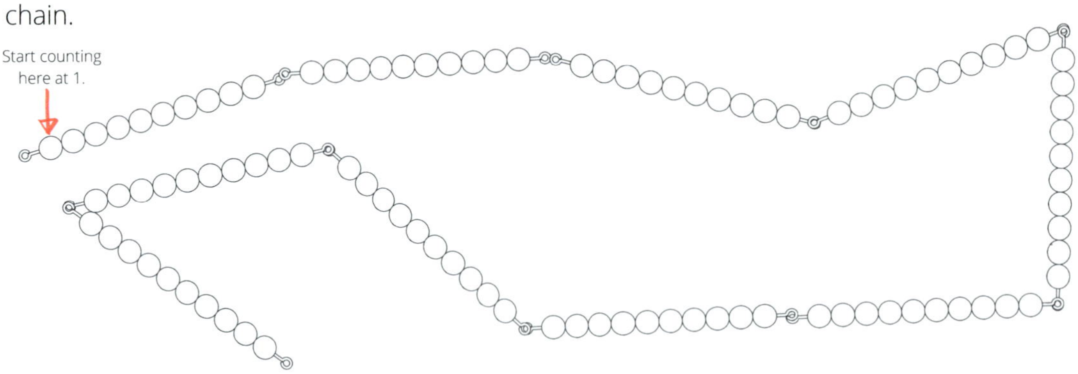

1 = red **22 = blue** **47 = green** **68 = yellow** **94 = pink**

19

M T W Th F S Su

Draw a line from the number to its spot in the one hundred board.

75

50

6

34

61

82

13

57

98

29

1	2	3	4	5		7	8	9	10
11	12		14	15	16	17	18	19	20
21	22	23	24	25	26	27	28		30
31	32	33		35	36	37	38	39	40
41	42	43	44	45	46	47	48	49	
51	52	53	54	55	56		58	59	60
	62	63	64	65	66	67	68	69	70
71	72	73	74		76	77	78	79	80
81		83	84	85	86	87	88	89	90
91	92	93	94	95	96	97		99	100

20

M T W Th F S Su

Color the short bead chains the proper color or "dab" them with paint using a cotton swab. Write the numbers on the bead chain arrows.

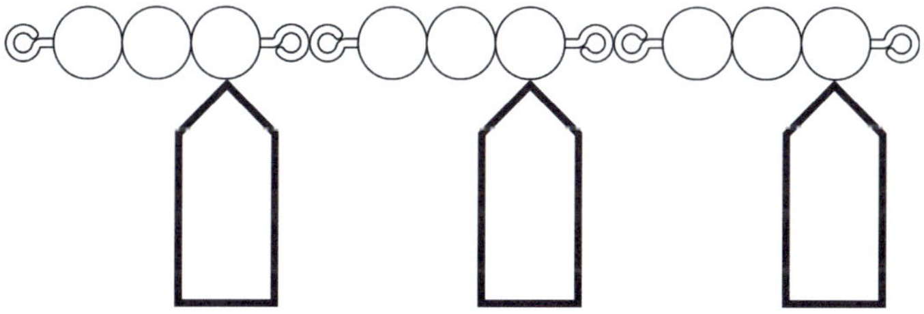

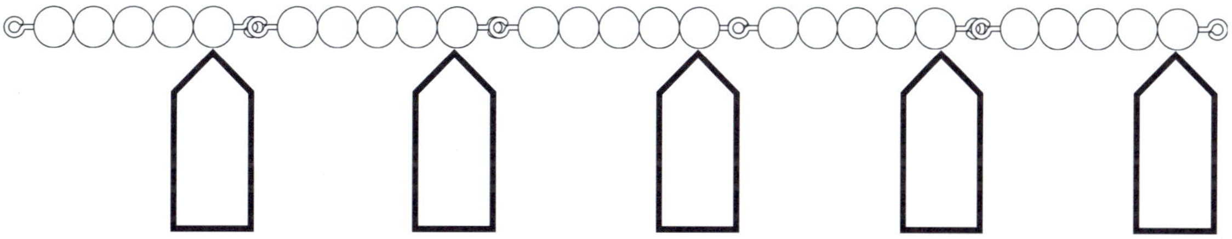

21

M T W Th F S Su

Color each bead square and complete the statements.

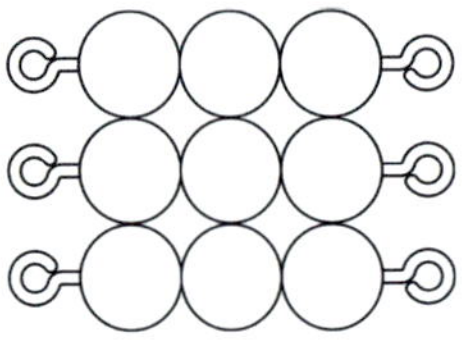

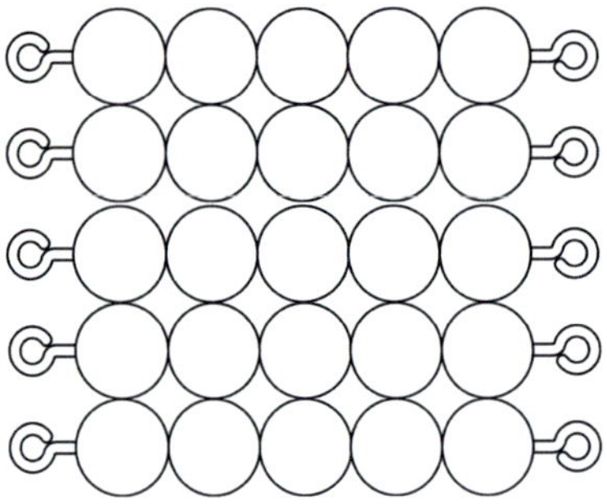

The square of ___ is _____.   The square of ___ is _____.

22

M T W Th F S Su

How many  are in  ?__________

How many 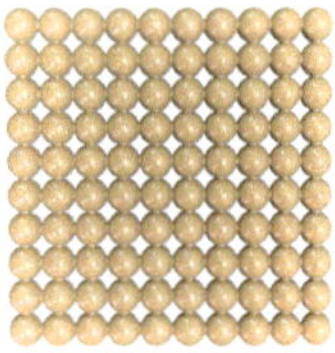 are in 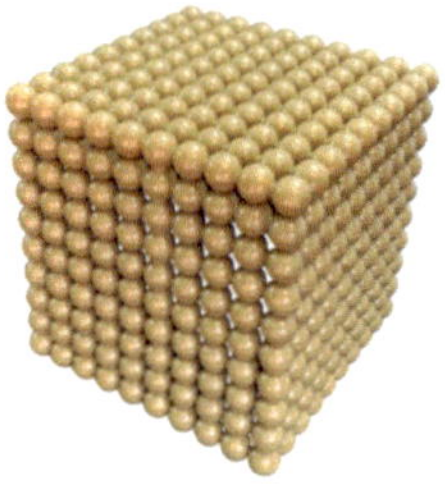 ?________

23

M T W Th F S Su

Write the number on the line. Review the 1000 Chain Lesson if necessary.

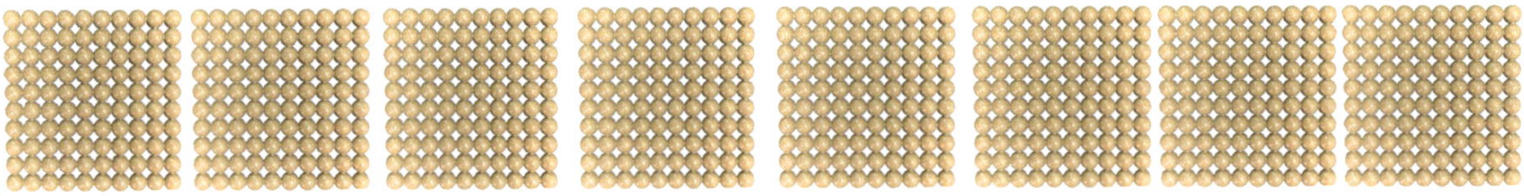

There are ____________ beads.

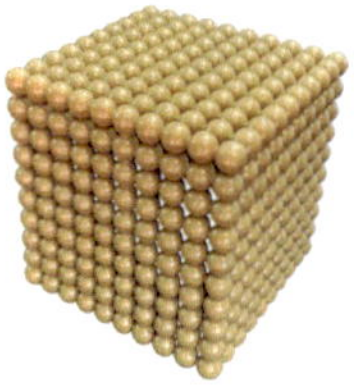

There are ____________ beads.

24

M T W Th F S Su

Draw a line from the number to the dot below the part of the mat where it belongs.
Then write the proper numeral in each white box.

1000	100	10	1

25

M T W Th F S Su

Draw a green line around a group of **five units.**
Draw a blue line around a group of **four tens.**
Draw a red line around a group of **two hundreds.**
Draw a green line around a group of **three thousands.**

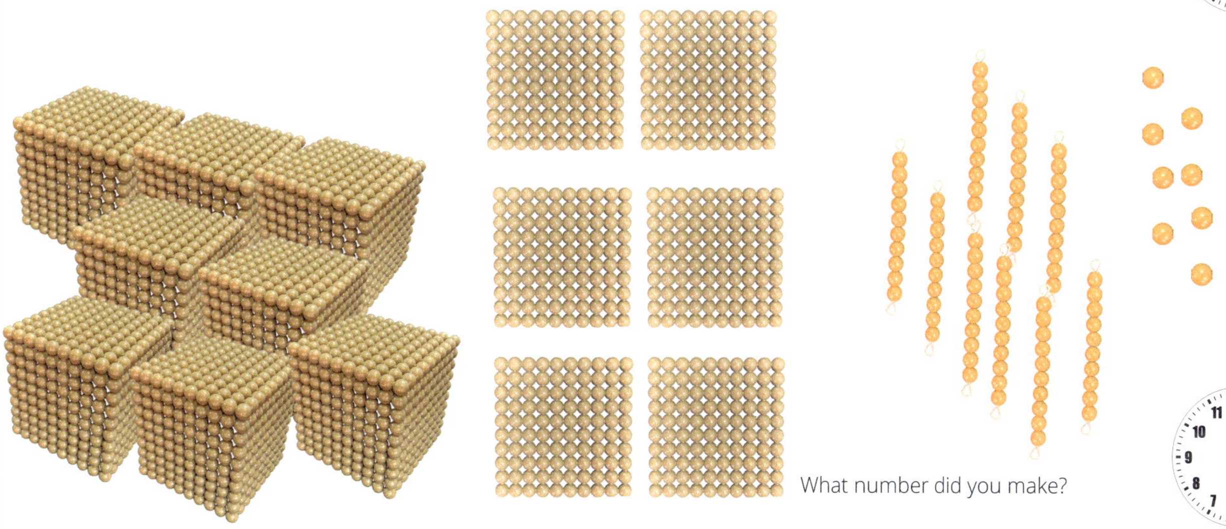

What number did you make?

26

M T W Th F S Su

Circle or shade in the corresponding decimal cards for this amount. Then write the number here:

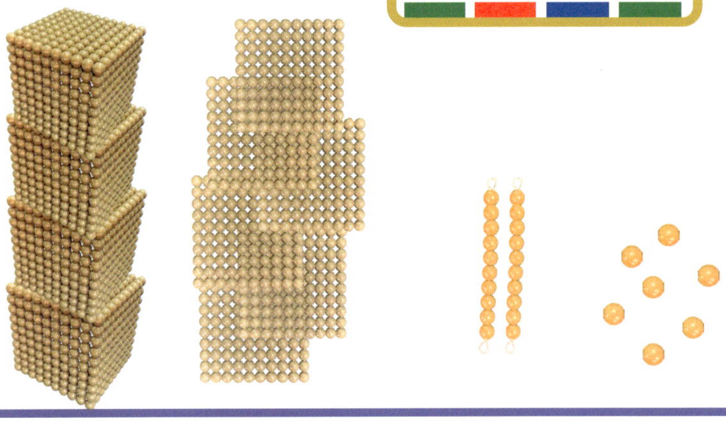

1000	100	10	1
2000	200	20	2
3000	300	30	3
4000	400	40	4
5000	500	50	5
6000	600	60	6
7000	700	70	7
8000	800	80	8
9000	900	90	9

27

M T W Th F S Su

Using the decimal cards as a control and guide, write each symbol in the boxes.

five units

three tens

nine thousands

eight tens

seven hundreds

1000	100	10	1
2000	200	20	2
3000	300	30	3
4000	400	40	4
5000	500	50	5
6000	600	60	6
7000	700	70	7
8000	800	80	8
9000	900	90	9

28

M T W Th F S Su

Color the corresponding amount of beads.

8 3 4 5

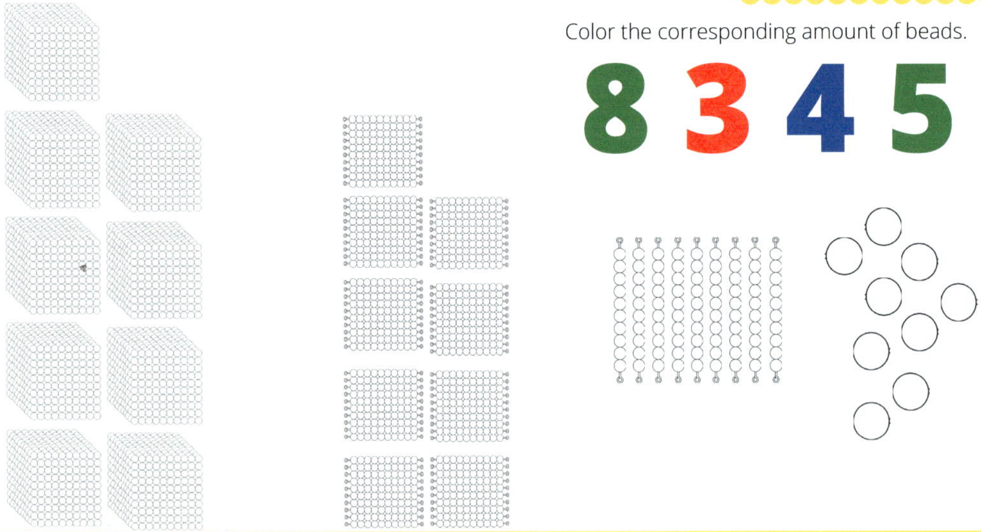

29

M T W Th F S Su

Using the simple drawing method for the Golden Beads, draw out the problem on the addition mat. When you move the beads "down" under the red addition line, redraw them in the lower section and cross them out from the upper sections.Write the numeral answer in the bottom left box.

	1245	1000	200	40	5
+	1163	1000	100	60	3

30

M T W Th F S Su

1. Write the numerals in the left column for the minuend and the subtrahend.
2. Cross out the beads from the top row (the minuend) that have been "moved down" to the middle row (the subtrahend).
3. "Move down" by drawing and then crossing out, the remainder from the minuend in the top row to the bottom row. The beads that are left in the bottom row is your remainder.
4. Write the numeral representing the remainder in the bottom left box.

*At the end of the problem, all of the beads in the top row should be crossed out with an "X."

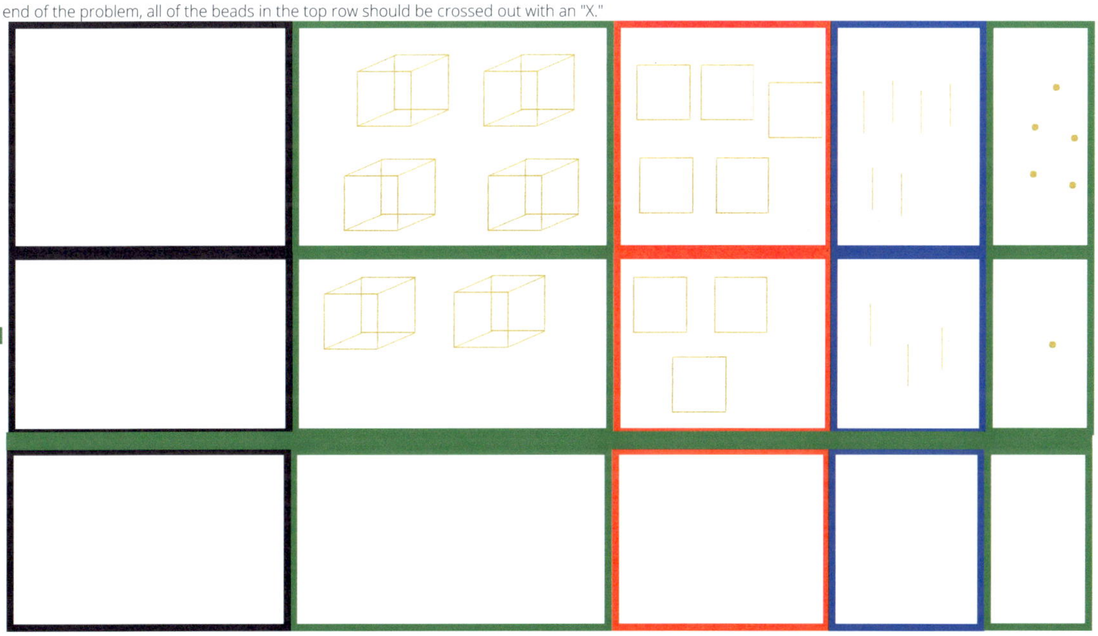

Looping Review Exercises

31

M T W Th F S Su

Draw a line from the numeral to the corresponding number rod.

1
2
3
4
5
6
7
8
9
10

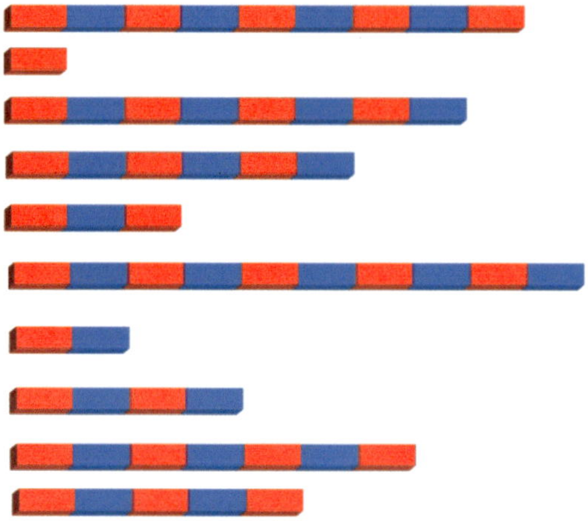

32

M T W Th F S Su

Draw counters under each numeral card, making sure to follow the pattern in the example to the right, with pairs matched and the last one by itself in the middle if it doesn't have a pair.

Then, circle whether the number is even or odd. Remind your child that even numbers all have pairs, and odd numbers have one "odd" one out.

HINT: You may want to use an erasable pencil in case your child decides to "move" the counter after it has been placed.

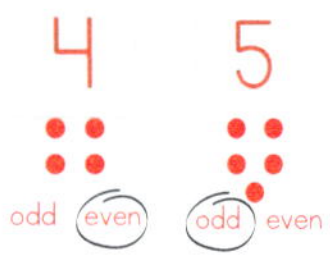

3 | 6 | 7 | 10

odd even | odd even | odd even | odd even

33

M T W Th F S Su

Write the numeral representing each number rod. Then determine the sum and write the numeral answer.

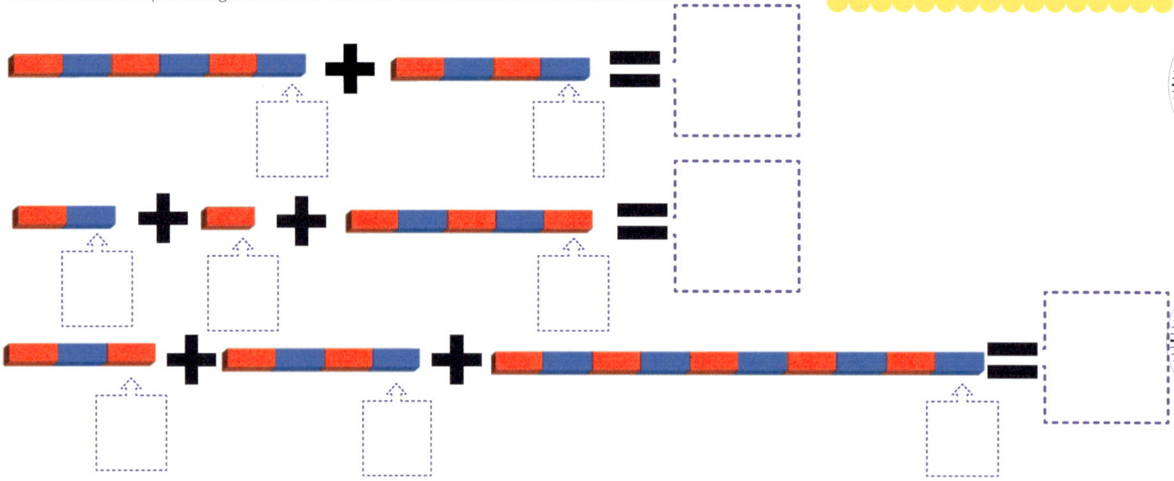

34

M T W Th F S Su

Use colored pencils to draw the indicated number rod, then circle the number of the rod you drew.

Draw the number rod that is one more than this one.

I drew the rod of: 1 2 3 4 5 6 7 8 9 10

Draw the number rod that is one less than this one.

I drew the rod of: 1 2 3 4 5 6 7 8 9 10

Draw the number rod that is two less than this one.

I drew the rod of: 1 2 3 4 5 6 7 8 9 10

35

M T W Th F S Su

Write the correct symbol in the circle.

Greater than >

Less than <

 ◯ 

 ◯ 

 ◯ 

6 ◯ 3

4 ◯ 10

Hint: The alligator eats the greater number. We call him the "greater gator."

Looping Review Exercises

M T W Th F S Su

36

Draw the rod longer until you get to ten. How many did you add? Write that number to the right in the circle.

I added:

M T W Th F S Su

37

Subtract from Ten. The top number rod is the Rod of Ten. Take away the rod under the rod of ten to the right by putting an "X" through it. Circle the rod that is left. Then write the subtraction problem. (Example: 10-2 = 8.)

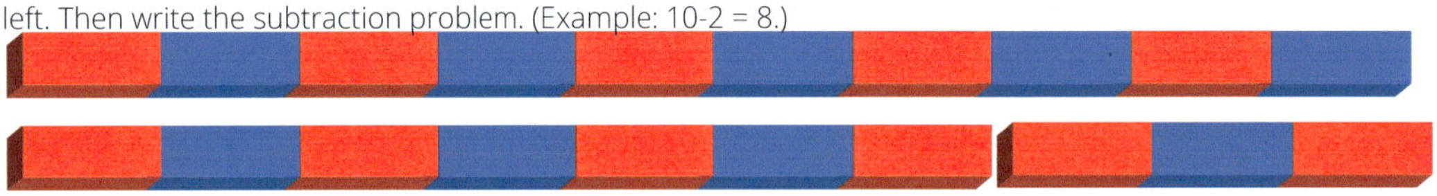

10 - ________ **=** ________

the top rod — the rod you crossed out — the rod that you circled

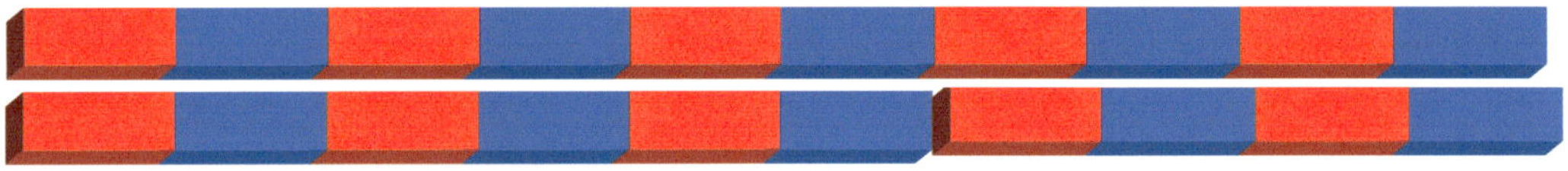

10 - ________ **=** ________

the top rod — the rod you crossed out — the rod that you circled

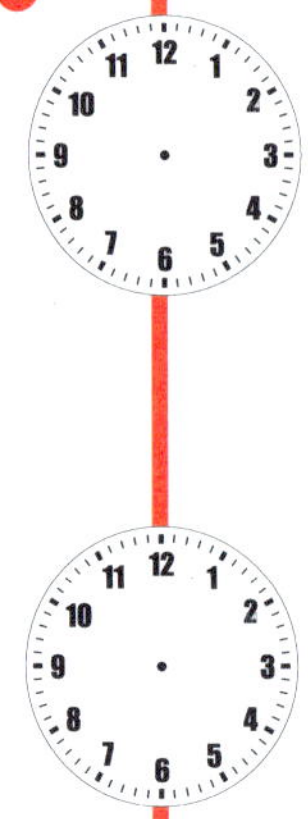

M T W Th F S Su

38

Dab and Count. Invite your child to use paints to paint the bead stair the proper colors by "dabbing" a cotton swab into paint and then once onto each bead. Use a control picture, if needed. Encourage your child to count aloud while dabbing.

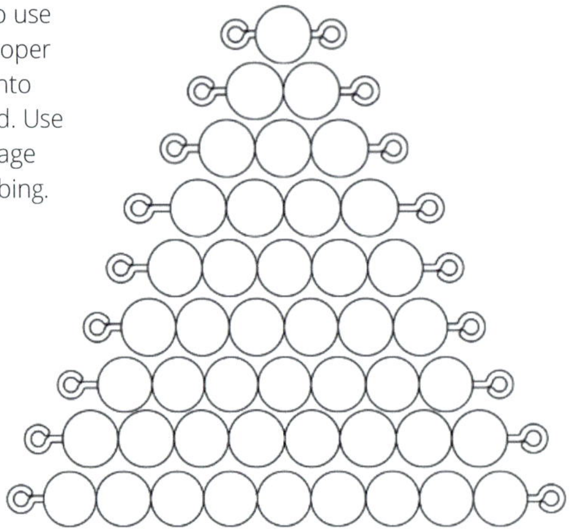

39

M T W Th F S Su

Write the sum as a numeral in the box.

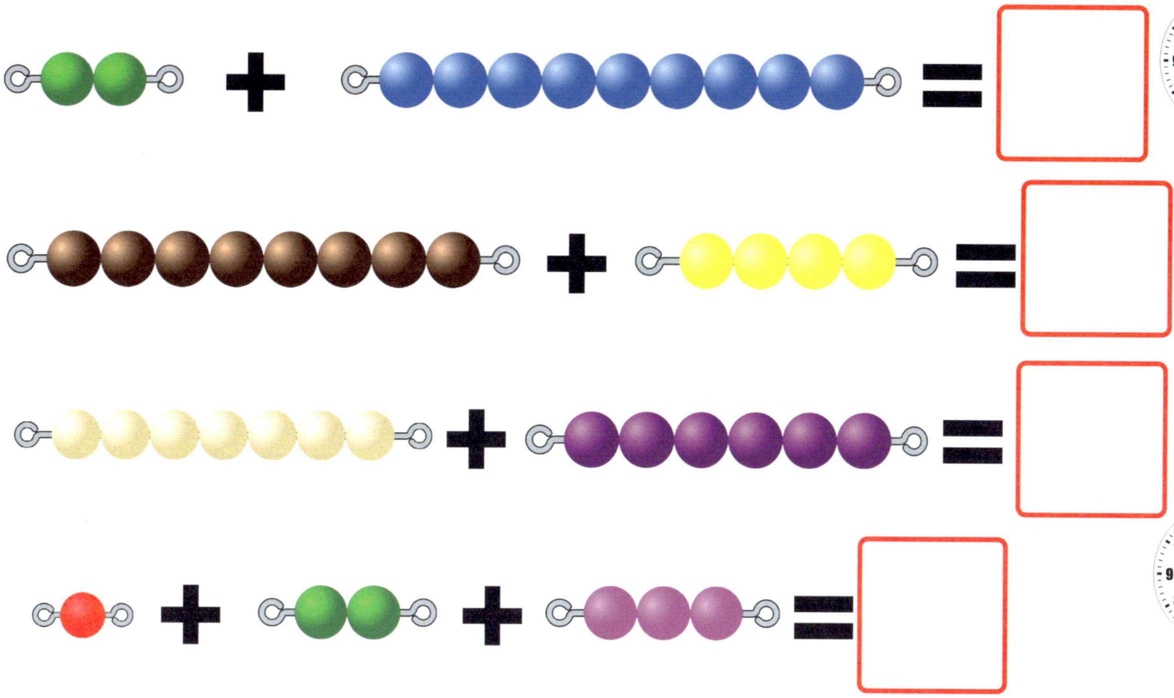

40

M T W Th F S Su

Make Ten Snake Game (without cutting and pasting)

Count to ten. Draw a pencil mark after the tenth bead. Cross out the ten beads you counted and draw a golden bead ten-bar to start your new snake. Keep counting from your pencil mark until you get to 10. Draw a pencil mark after the tenth bead. Cross out the ten beads you counted and draw another golden bead ten-bar on your new snake. If there are less than ten beads at the end, draw a black and white bead bar.

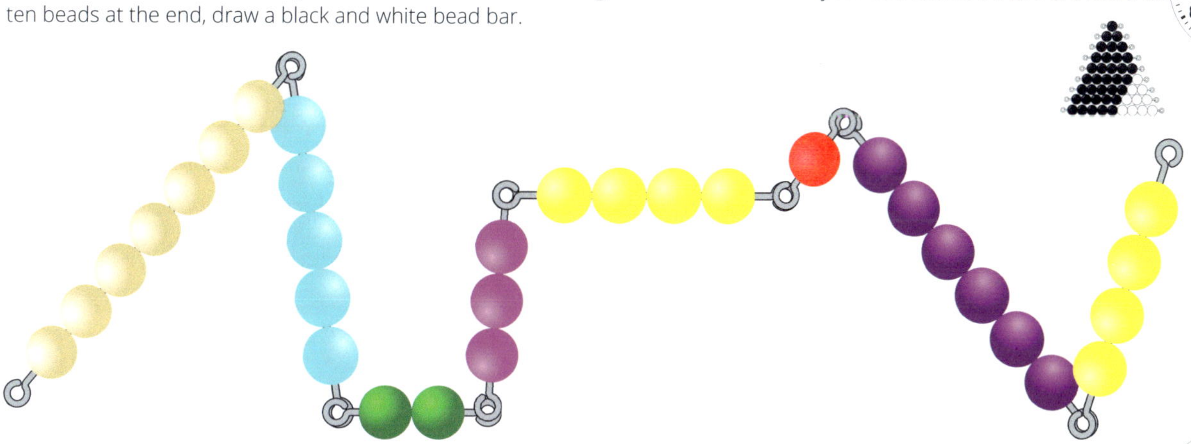

41

M T W Th F S Su

Draw the colored bead bar to the right of the ten-bar that you need to make the teen number.

Example:

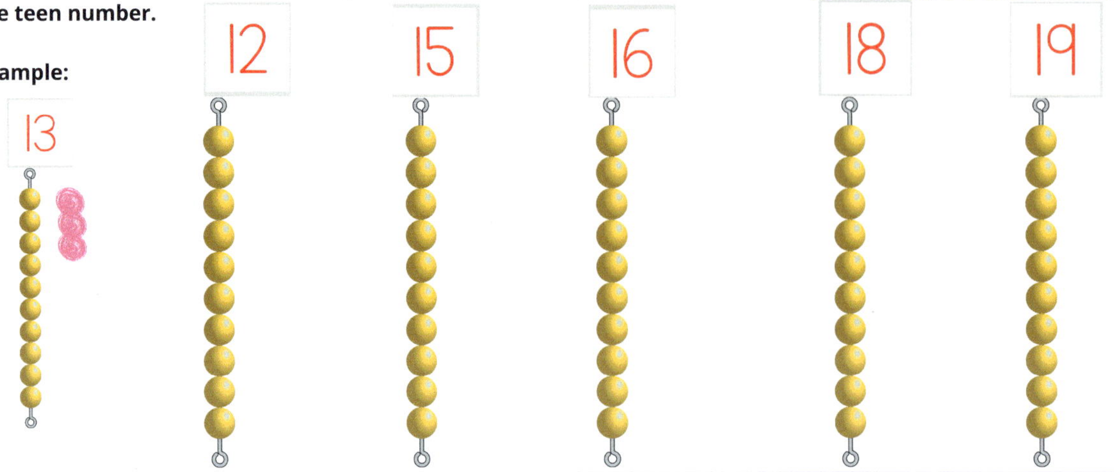

42

M T W Th F S Su

Draw a line from the numeral tile to the space it belongs on the teen board for numbers 16-19.

Then, write in the numerals needed to make 16-19 on the other board.

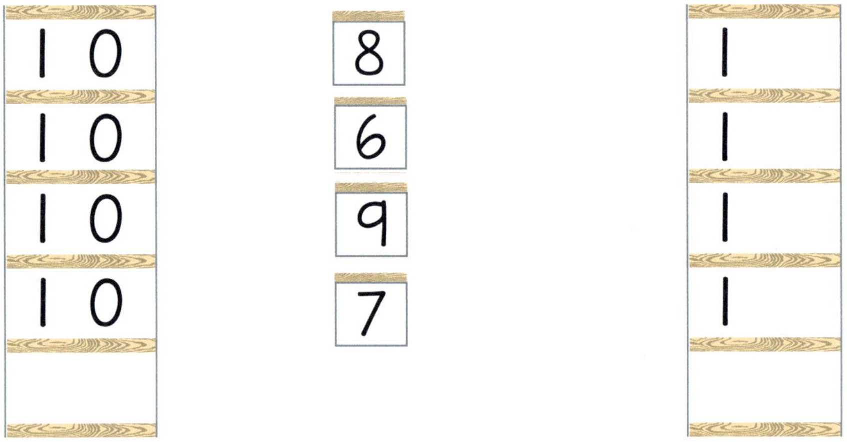

43

M T W Th F S Su

Draw the colored bead bars their proper colors, in order from left to right, for 11-19, and write in the numerals on the top tiles.

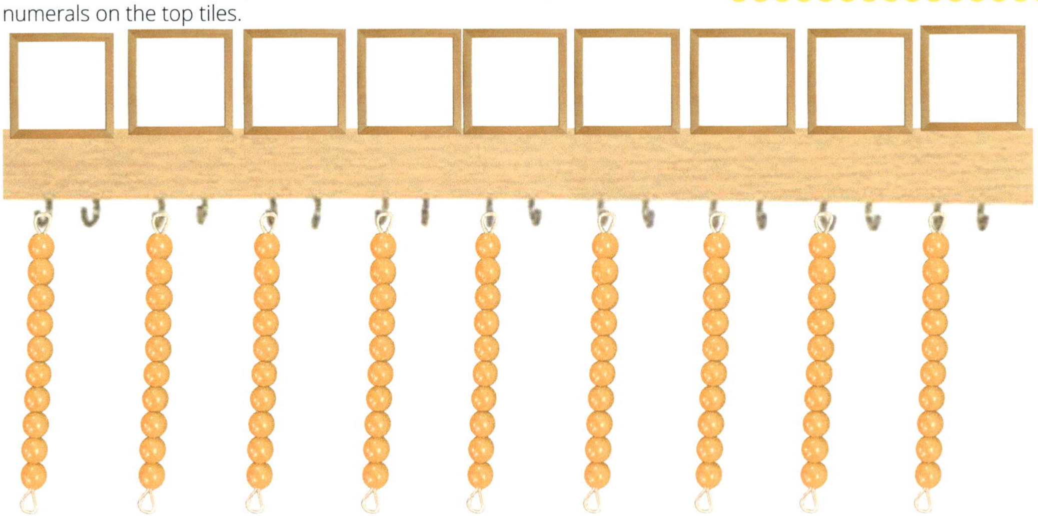

44

M T W Th F S Su

What teen value is made from combining the two number rods? Write the teen numeral in the box.

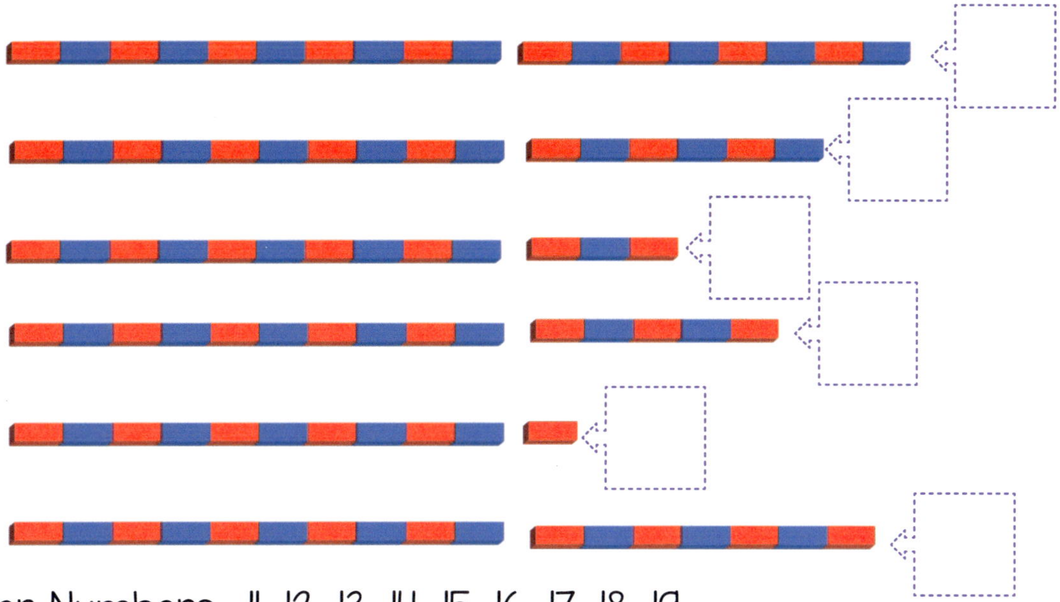

Teen Numbers: 11 12 13 14 15 16 17 18 19

45

M T W Th F S Su

How many tens are represented in each picture?
Write the numeral in the box. (Review numerals at the bottom of the page.)
(Two tens is written as 20.)

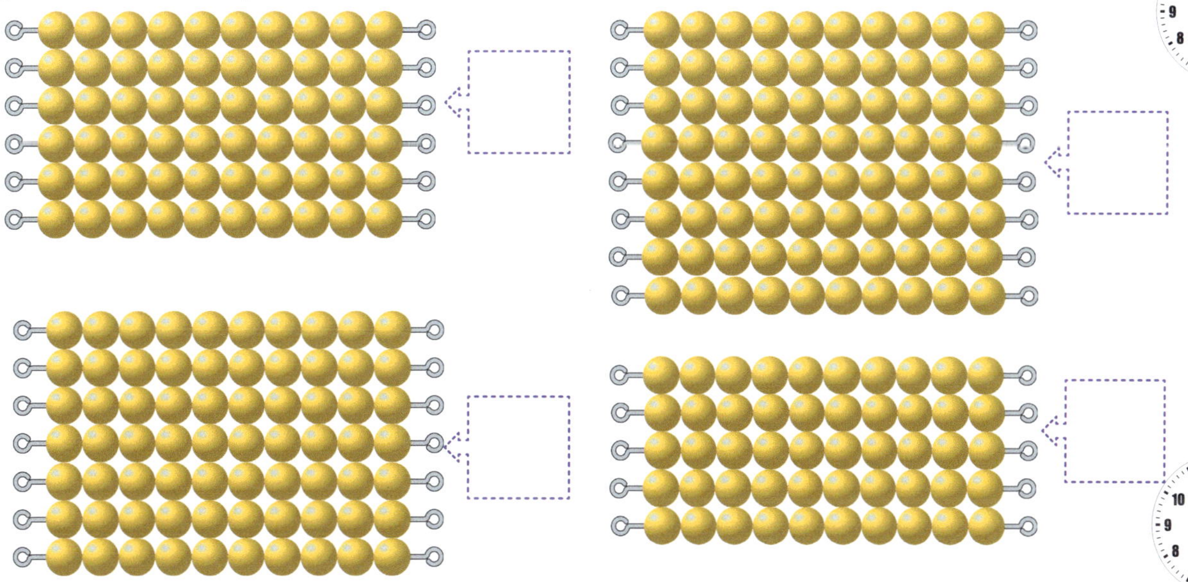

Tens: 10 20 30 40 50 60 70 80 90

46

M T W Th F S Su

Write the numbers 10-90 on the ten boards. Remember, the last spot is blank!

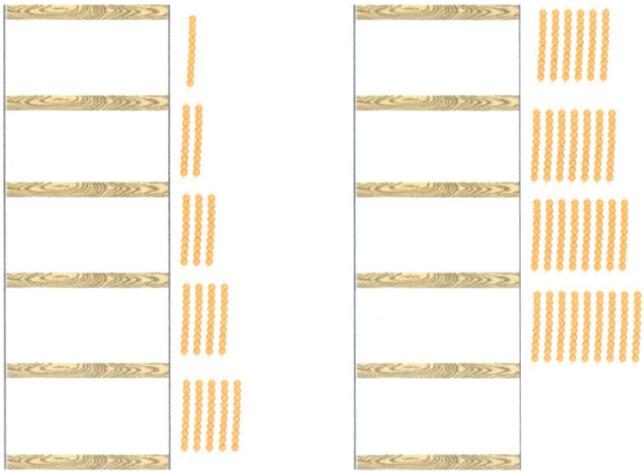

47

M T W Th F S Su

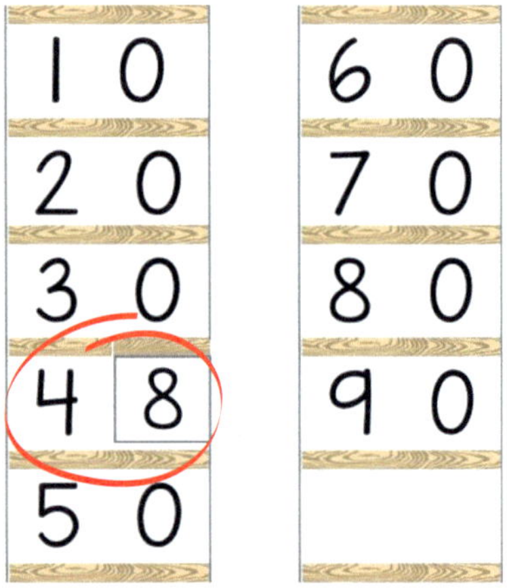

Draw the Golden Beads represented by the circled numeral. Use a line for ten-bars and dots for unit beads.

10 1

48

M T W Th F S Su

Color the beads along the 100 Chain according to the key below the chain.

Start counting here at 1.

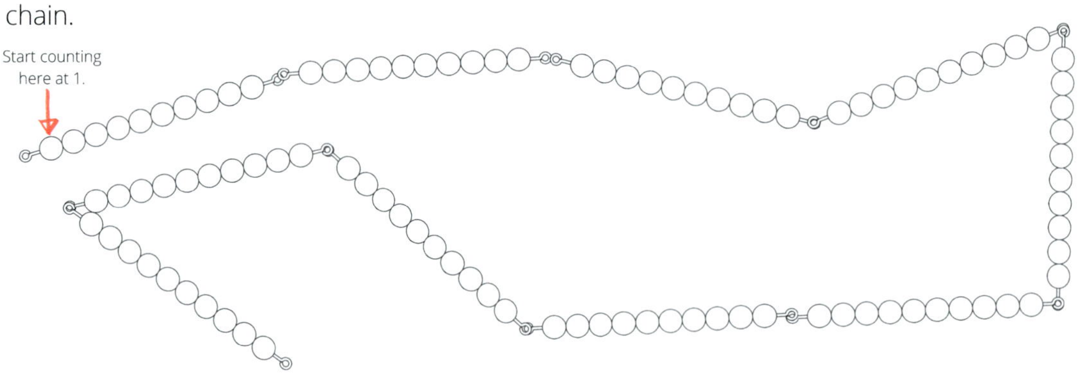

10 = red **39 = blue** **51 = green** **77 = yellow** **100 = pink**

49

M T W Th F S Su

Draw a line from the number to its spot in the one hundred board.

30
100
50
55
16
31
83
1
71
68

	2	3	4	5	6	7	8	9	10
11	12	13	14	15		17	18	19	20
21	22	23	24	25	26	27	28	29	
	32	33	34	35	36	37	38	39	40
41	42	43	44	45	46	47	48	49	
51	52	53	54		56	57	58	59	60
61	62	63	64	65	66	67		69	70
	72	73	74	75	76	77	78	79	80
81	82		84	85	86	87	88	89	90
91	92	93	94	95	96	97	98	99	

50

M T W Th F S Su

Color the short bead chains the proper colors. Using a cotton ball and paint to "dab" the beads is a fun alternative to coloring with pencil! Write the numbers on the bead chain arrows.

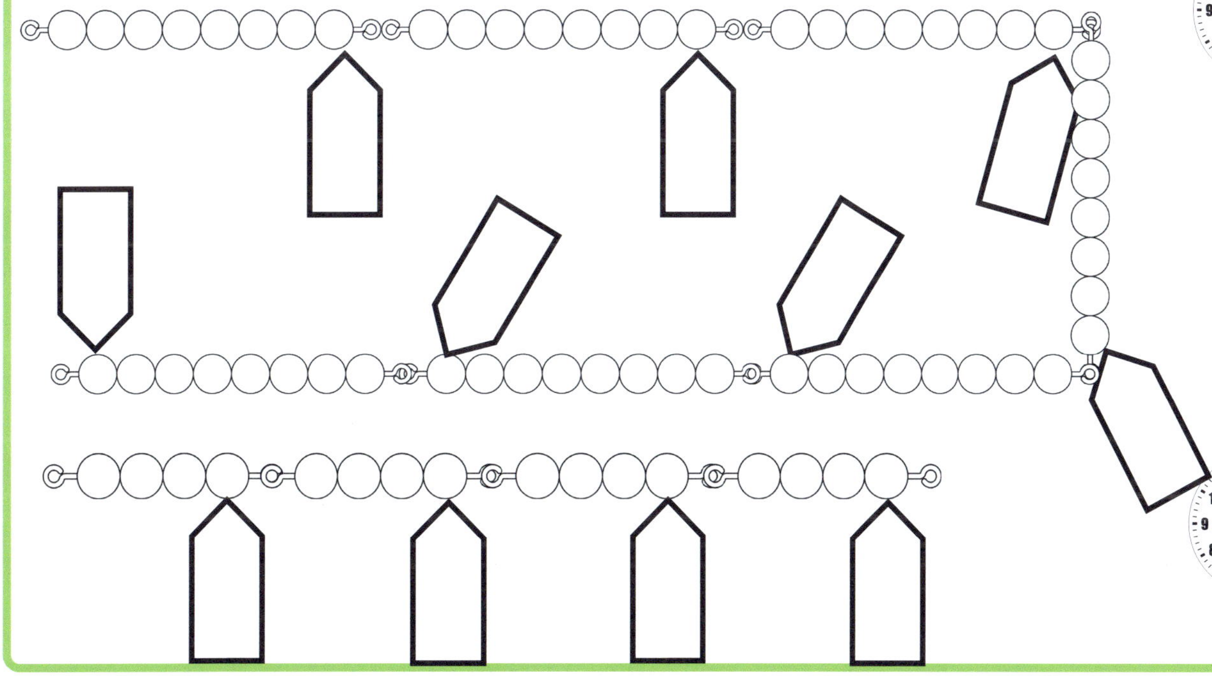

51

M T W Th F S Su

Color each bead square and complete the statements.

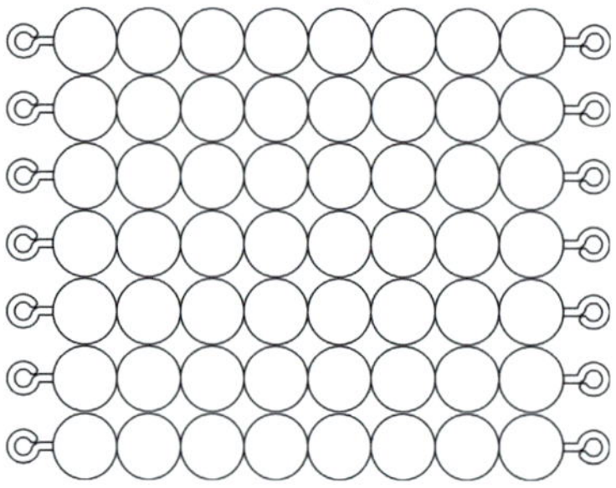

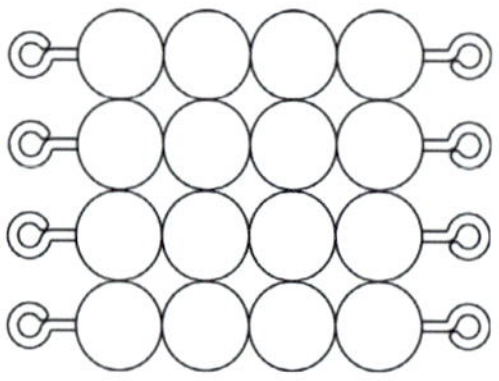

The square of ___ is _____. The square of ___ is _____.

52

M T W Th F S Su

How many  are in 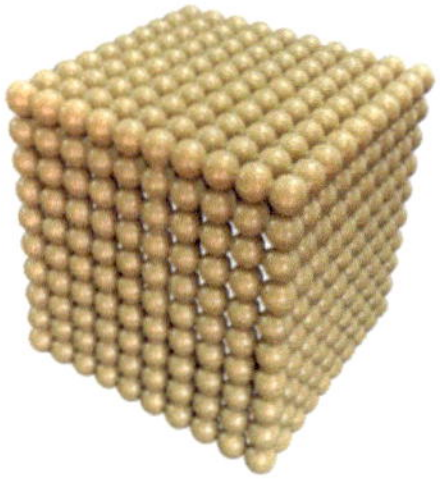 ?________

How many  are in  ?________

53

M T W Th F S Su

Draw the number of hundred squares that are found in the thousand cube.

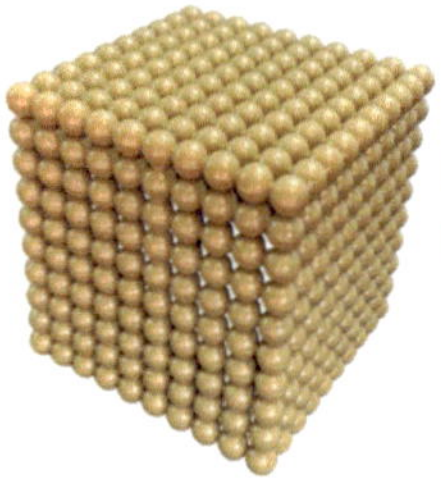 =

54

M T W Th F S Su

Draw a line from the beads to the dot below the part of the mat where it belongs. Say the amounts out loud.

55

M T W Th F S Su

Draw a green line around a group of **eight units.**
Draw a blue line around a group of **five tens.**
Draw a red line around a group of **six hundreds.**
Draw a green line around a group of **seven thousands.**

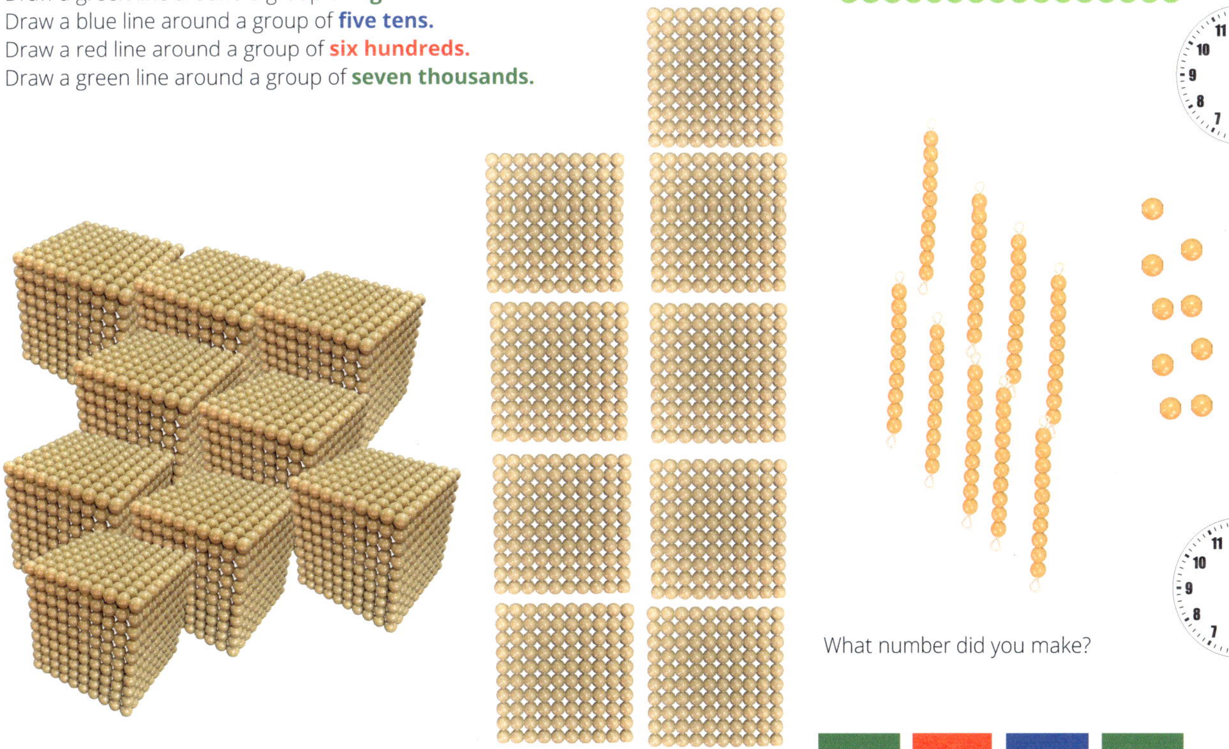

What number did you make?

Looping Review Exercises

56

M T W Th F S Su

Circle or shade in the corresponding decimal cards for this amount. Then write the number here:

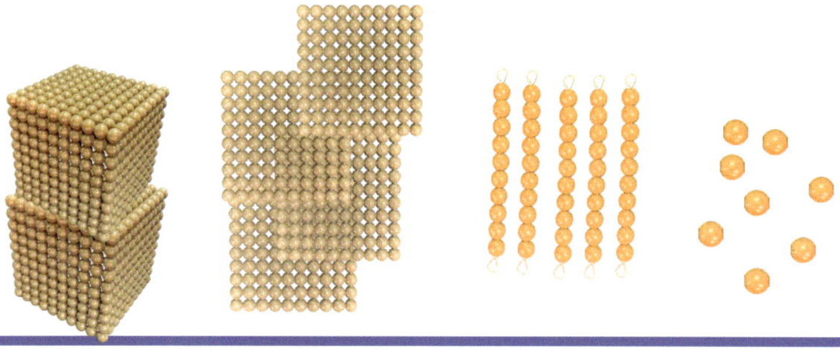

1000	100	10	1
2000	200	20	2
3000	300	30	3
4000	400	40	4
5000	500	50	5
6000	600	60	6
7000	700	70	7
8000	800	80	8
9000	900	90	9

57

M T W Th F S Su

Using the decimal cards as a control and guide, write each symbol in the boxes.

seven thousands

two tens

four hundreds

nine units

six hundreds

1000	100	10	1
2000	200	20	2
3000	300	30	3
4000	400	40	4
5000	500	50	5
6000	600	60	6
7000	700	70	7
8000	800	80	8
9000	900	90	9

58

M T W Th F S Su

Color the corresponding amount of beads.

4 7 9 2

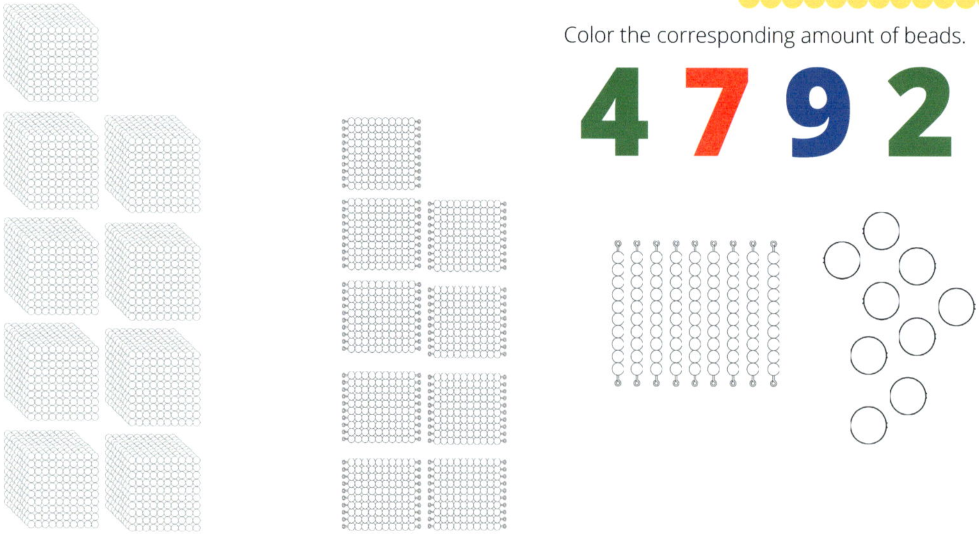

59

M T W Th F S Su

Using the simple drawing method for the Golden Beads, draw out the problem on the addition mat. When you move the beads "down" under the red addition line, redraw them in the lower section and cross them out from the upper sections.Write the numeral answer in the bottom left box.

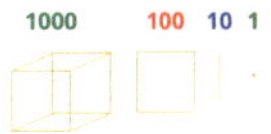

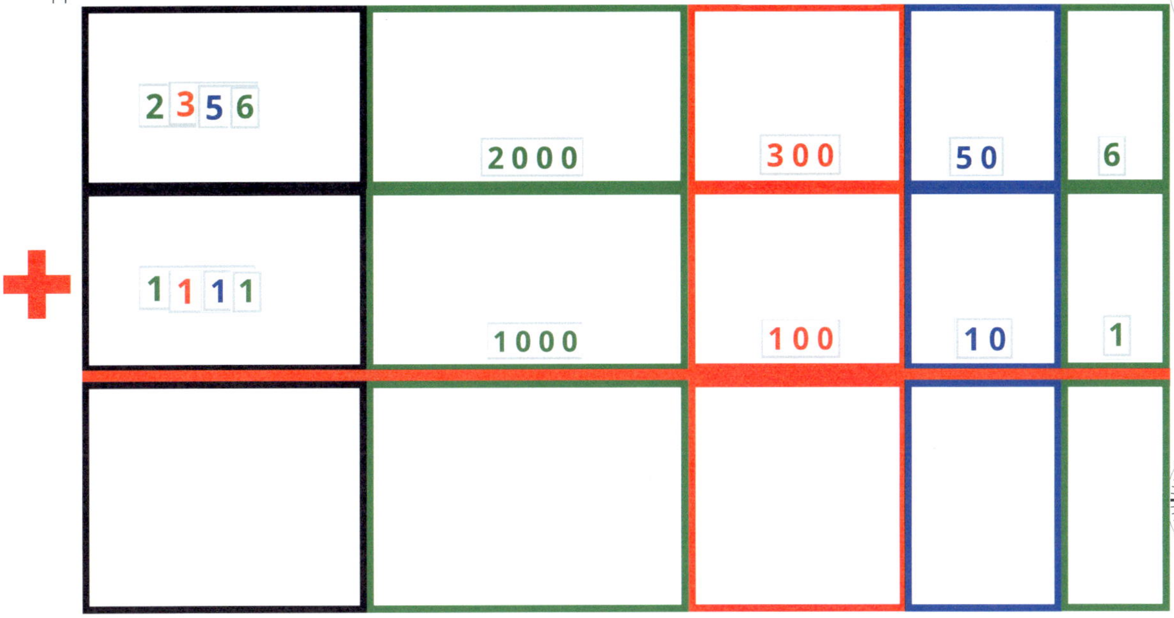

60

M T W Th F S Su

1. Write the numerals in the left column for the minuend and the subtrahend.
2. Cross out the beads from the top row (the minuend) that have been "moved down" to the middle row (the subtrahend).
3. "Move down" by drawing and then crossing out, the remainder from the minuend in the top row to the bottom row. The beads that are left in the bottom row is your remainder.
4. Write the numeral representing the remainder in the bottom left box.

*At the end of the problem, all of the beads in the top row should be crossed out with an "X."

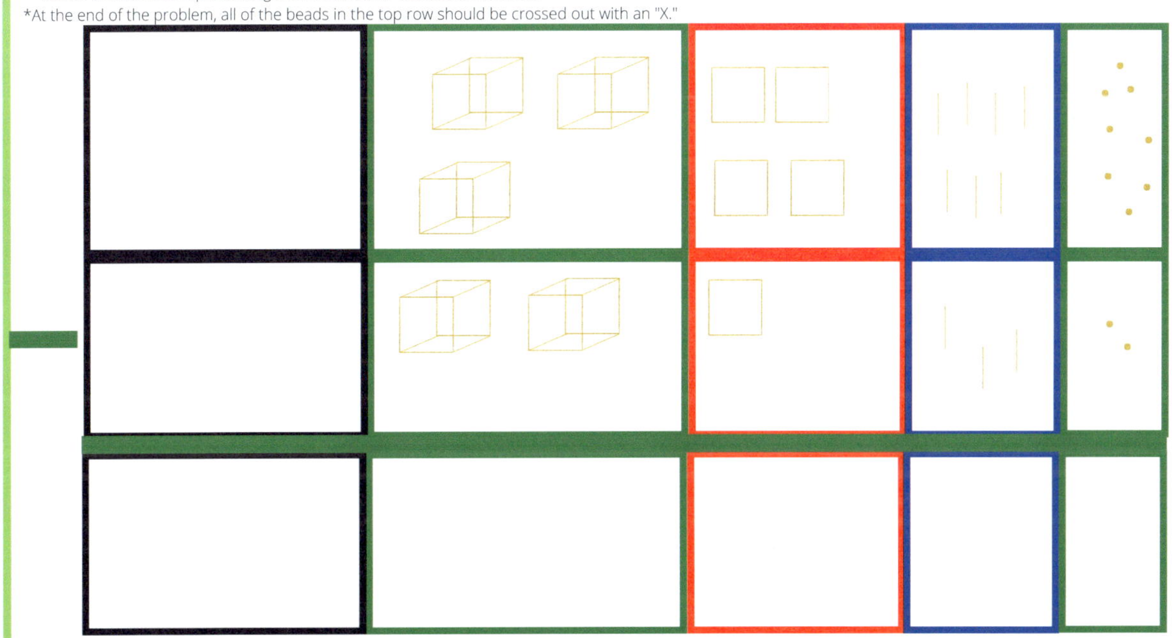

Book 2 Completed

Name:__________________________

Date Completed:____________

I can add, subtract, multiply, and divide four digit numbers!

Dear __________________,

You did it! You worked hard and completed your **Montessori Math Workbook, Book 2!** I am most proud of you for ______________________________________.
I love how you ______________________________
____.
You have proved you are capable of
____!
Most of all, I want you to know that *I love you*, and we are on this journey together, every step of the way.

Love, ________________________

Made in the USA
Las Vegas, NV
04 November 2026

57431707R00133